SWB—A Modified Thornthwaite-Mather Soil-Water-Balance Code for Estimating Groundwater Recharge

Chapter 31 of
Section A, Groundwater, of
Book 6, Modeling Techniques

Groundwater Resources Program

Techniques and Methods 6–A31

U.S. Department of the Interior
U.S. Geological Survey

U.S. Department of the Interior
KEN SALAZAR, Secretary

U.S. Geological Survey
Marcia K. McNutt, Director

U.S. Geological Survey, Reston, Virginia: 2010

For more information on the USGS—the Federal source for science about the Earth, its natural and living resources, natural hazards, and the environment, visit http://www.usgs.gov or call 1-888-ASK-USGS

For an overview of USGS information products, including maps, imagery, and publications, visit http://www.usgs.gov/pubprod

To order this and other USGS information products, visit http://store.usgs.gov

Suggested citation:
Westenbroek, S.M., Kelson, V.A., Dripps, W.R., Hunt, R.J., and Bradbury, K.R., 2010, SWB—A modified Thornthwaite-Mather Soil-Water-Balance code for estimating groundwater recharge: U.S. Geological Survey Techniques and Methods 6–A31, 60 p.

Preface

Performance of this computer program has been tested and verified for several test cases; however, future applications of the program could reveal errors that were not detected in the test cases. Users are requested to notify the U.S. Geological Survey (USGS) if errors are found in the documentation report or in the computer program.

Correspondence regarding the report or program should be sent to

USGS Wisconsin Water Science Center
8505 Research Way
Middleton, WI   53562–3581
Attention: Stephen M. Westenbroek
Email: smwesten@usgs.gov

Although the computer program has been used by the USGS, no warranty, expressed or implied, is made by the USGS or the United States Government as to the accuracy and functionality of the program and related program material. Nor shall the fact of distribution constitute any such warranty, and no responsibility is assumed by the USGS in connection therewith.

The Soil-Water-Balance code and other groundwater programs are available for downloading from the USGS at the following world wide web address: *http://water.usgs.gov/software/ground_water.html*.

Acknowledgments

The U.S. Geological Survey's Groundwater Resources Program funded this work. We appreciate the comments and input provided by Peter Schoephoester and David Hart (Wisconsin Geologic and Natural History Survey) and David Saad (U.S. Geological Survey). We also appreciate the feedback from several individuals who tested and worked with earlier versions of the SWB code: Tim Brown (Barr Engineering), Todd Rayne (Hamilton College), and Samantha Lax (Wittman Hydro Plannning Associates). Lastly, we wish to thank Daniel Feinstein (U.S. Geological Survey), whose ideas led to development of the code allowing spatially interpolated data to be used as the driver for SWB model simulations.

Contents

Preface iii

Acknowledgments iv

Abstract 1

Introduction 1

    Background 1

    Purpose and Scope 2

Model Description 2

    Model Theory 2

    Data Requirements for Application of the SWB Model 5

    Summary of Major Differences Between the Original and Current SWB Codes 5

    Extensions and Additional Capabilities of the SWB Code 7

    Model Limitations and Assumptions 8

Use of the SWB Model 9

    Required Directory Structure 9

    Input Files 10

        Tabular Files 10

        Gridded Files 10

        Lookup Tables 14

    Output Files 17

        Tabular Files 17

        Grid and Image Files 19

    Program Options and the Control File 21

        Required Control-File Entries 21

        Optional Control-File Entries 26

Test Problems 31

    Test Case 1—Black Earth Creek, Dane and Iowa Counties, Wisconsin 31

        Input Tables and Grids 31

        Simulation Details 31

        Simulation Results 31

    Test Case 2—Lake Michigan Basin 40

        Input Tables and Grids 40

        Simulation Details 46

        Simulation Results 46

        Model Parameter Sensitivity 46

Summary and Conclusions 50

References Cited 50

Appendix 1: SWB Module Description 54

    Main Program and Module 54

    Support Modules 54

    Process Modules 54

        Evapotranspiration 54

        Runoff 54

        Soil Moisture 54

Appendix 2: Preparation of Gridded Climatological Data for SWB 55
    Climatological Data Source 55
    Performing Interpolations 57

Figures

1–3.   Diagrams showing:
    1.   Interaction between Soil-Water-Balance code and data 6
    2.   Definition of closed depression based on flow-direction-grid inputs 7
    3.   Required directory structure 9
  4.   Example of Arc ASCII grid input-file format...11
  5.   Diagram showing numerical definition of D8 flow directions 13
  6.   Graph showing soil-moisture-retention table..19
7–20.   Maps showing:
    7.   Land-cover classification for the Black Earth Creek test case 32
    8.   Hydrologic soil groups for the Black Earth Creek test case....................................33
    9.   D8 flow-direction grid for the Black Earth Creek test case....................................34
    10.   Available water capacity (AWC) for soils in the Black Earth Creek
        test case...35
    11.   Comparison of Fortran 95 and Visual Basic versions of the SWB code—
        net infiltration, 1999 36
    12.   Comparison of Fortran 95 and Visual Basic versions of the SWB code—
        actual evapotranspiration, 1999 37
    13.   Comparison of Fortran 95 and Visual Basic versions of the SWB code—
        ending soil moisture, 1999..38
    14.   Comparison of Fortran 95 and Visual Basic versions of the SWB code—
        recharge, 1999..39
    15.   Land-cover classification for the Lake Michigan Basin test case...........................41
    16.   Hydrologic soil groups for the Lake Michigan Basin test case 42
    17.   D8 flow-direction grid for the Lake Michigan Basin test case 43
    18.   Available water capacity (AWC) for soils in the Lake Michigan Basin
        test case...44
    19.   Mean annual precipitation (1990–2000) for the Lake Michigan Basin
        test case...45
    20.   Example recharge (1990–2000) for the Lake Michigan Basin test case..................47
21–22.   Plots showing:
    21.   Comparison of SWB-estimated recharge to Q90-estimated recharge for
        basins with drainage areas greater than 50 square miles 48
    22.   Relative parameter sensitivities for the Lake Michigan Basin SWB model...........49

Appendix Figures

2–1. Example of the output from the thin plate spline technique in the fields library of the R statistical package 58

Tables

1. Definition of antecedent runoff conditions used in the SWB code......................................3
2. Data requirements for application of the SWB model 5
3. Example of the data elements contained in the tabular climate data file........................10
4. Data requirements for various potential evapotranspiration methods 11
5. Modified Anderson Level II land-use classification scheme..12
6. Infiltration rates for Natural Resources Conservation Service hydrologic soil groups...12
7. Estimated available water capacities for various soil-texture groups.............................13
8. Data elements contained in the land-use lookup table..14
9. Annotated example land-use lookup table..15
10. Provisional water-holding capacities with different combinations of soil and vegetation 18
11. Description of available output variables...20

Conversion Factors

Multiply	By	To obtain
	Length	
inch (in.)	2.54	centimeter (cm)
inch (in.)	25.4	millimeter (mm)
meter (m)	3.281	foot (ft)
foot (ft)	.3048	meter (m)
mile (mi)	1.609	kilometer (km)
	Area	
square mile (mi^2)	259.0	hectare (ha)
square mile (mi^2)	2.590	square kilometer (km^2)
	Volume	
cubic foot (ft^3)	.02832	cubic meter (m^3)
acre-foot (acre-ft)	1,233	cubic meter (m^3)
	Velocity	
meter per second (m/s)	3.281	foot per second (ft/s)
millimeter per year (mm/yr)	.03937	inch per year (in/yr)
inch per year (in/yr)	25.4	millimeter per year (mm/yr)
	Flow rate	
acre-foot per day (acre-ft/d)	.01427	cubic meter per second (m^3/s)
acre-foot per year (acre-ft/yr)	1,233	cubic meter per year (m^3/yr)
cubic foot per second (ft^3/s)	.02832	cubic meter per second (m^3/s)
cubic foot per day (ft^3/d)	.02832	cubic meter per day (m^3/d)

Temperature in degrees Celsius (°C) may be converted to degrees Fahrenheit (°F) as follows:
°F=(1.8×°C)+32

Temperature in degrees Fahrenheit (°F) may be converted to degrees Celsius (°C) as follows:
°C=(°F-32)/1.8.

SWB—A Modified Thornthwaite-Mather Soil-Water-Balance Code for Estimating Groundwater Recharge

By S.M. Westenbroek, V.A. Kelson[1], W.R. Dripps[2], R.J. Hunt, and K.R. Bradbury[3]

Abstract

A Soil-Water-Balance (SWB) computer code has been developed to calculate spatial and temporal variations in groundwater recharge. The SWB model calculates recharge by use of commonly available geographic information system (GIS) data layers in combination with tabular climatological data. The code is based on a modified Thornthwaite-Mather soil-water-balance approach, with components of the soil-water balance calculated at a daily timestep. Recharge calculations are made on a rectangular grid of computational elements that may be easily imported into a regional ground-water-flow model. Recharge estimates calculated by the code may be output as daily, monthly, or annual values.

Introduction

Accurate estimates of the spatial and temporal distribution of recharge are important for many types of hydrologic assessments, including those that concern water-quality protection, streamflow and riparian ecosystem management, aquifer replenishment, groundwater-flow modeling, and contaminant transport. These recharge estimates are often key to understanding the effects of development in urban, industrial, and agricultural regions. With increasing demand for hydrologic assessments in support of management decisions comes an increased need for practical methods to quantify recharge rates and delineate zones of similar recharge (Scanlon and others, 2002).

The Soil-Water-Balance code has been developed to allow estimates of recharge to be made quickly and easily. The code calculates components of the water balance at a daily timestep by means of a modified version of the Thornthwaite-Mather soil-water-balance approach (Thornthwaite, 1948; Thornthwaite and Mather, 1957). Data requirements include

several commonly available tabular and gridded data types: (1) precipitation and temperature, (2) land-use classification, (3) hydrologic soil group, (4) flow direction, and (5) soil-water capacity. The data and formats required are designed to take advantage of widely available GIS datasets and file structures.

Background

Groundwater recharge can vary greatly over time and space. Site-specific data, when available, are not applicable to regional-scale problems. Groundwater modelers often assume that a fraction of precipitation is converted to recharge, or they instead use recharge as a calibration parameter. In transient groundwater-modeling problems, use of a physically based, spatially variable recharge boundary condition has been found to improve model performance (Jyrkama and Sykes, 2007).

Numerical modeling is one technique sometimes used to supply a spatially varied, transient recharge boundary condition on a regional scale (Scanlon and others, 2002). Simple soil-water-balance models are a category of numerical models commonly applied to groundwater recharge estimation problems. There perhaps are hundreds of soil-water-balance models described in the literature. Many soil-water-balance models were developed in order to evaluate crop irrigation requirements and impacts (Kendy and others, 2003), crop yield prediction (Akinremi and others, 1996), and landfill cover design (Schroeder and others, 1994).

Similarly, there are many examples of groundwater recharge estimation by means of a soil-water balance. For example, the U.S. Environmental Protection Agency (U.S. EPA) HELP model, a soil-water-balance code used in landfill design (Schroeder and others, 1994), has been linked to commercial geographic information system (GIS) software (Jyrkama and others, 2002). WetSpass calculates long-term recharge by means of a soil-water-balance model within a commercial GIS software package (Batelaan and De Smedt, 2001). Finch (2001) describes a distributed daily soil-water-balance model, but does not specify the computing platform.

The SWB code described in this report was derived from work completed as part of a doctoral dissertation at the University of Wisconsin–Madison (Dripps, 2003; Dripps and

[1]Wittman Hydro Planning Associates, Bloomington, Indiana

[2]Furman University, Greenville, South Carolina

[3]Wisconsin Geologic and Natural History Survey, Madison, Wisconsin

Bradbury, 2007). This code was written in Visual Basic for Applications inside of a Microsoft Excel spreadsheet (Dripps, 2003).

Each of the examples listed above either requires proprietary software, is implemented in a proprietary language, is complex to set up and use, or is not distributed in the public domain. For these reasons, in 2006, the U.S. Geological Survey (USGS) translated the original soil-water-balance code from Visual Basic to modern Fortran 95.

Purpose and Scope

This report documents the Soil-Water-Balance (SWB) computer code, which is designed to calculate the spatial distribution of groundwater recharge over time using a gridded data structure. The SWB model should yield better results than can be obtained by assuming that a fraction of precipitation converts to recharge; conversely, the SWB model is much simpler and less time-intensive to apply than a fully coupled groundwater and surface-water model (Markstrom and others, 2008; Jyrkama and others, 2002).

The theoretical basis, data requirements for use, and limitations and assumptions relating to the SWB code are presented in this report. The requirements for application of the SWB code, including the directory structure and input files, the types of output available, and the various control-file options are described. In addition, two test cases are provided. The first test case (Black Earth Creek) confirms the numerical accuracy of the SWB code relative to the original Visual Basic code on which it is based. The second test case demonstrates the application of the SWB code to the Lake Michigan Basin, with a model domain covering about 116,180 mi^2.

Model Description

This section describes the theoretical basis, data requirements for use, and limitations and assumptions relating to the SWB code. For greater theoretical detail, the reader is directed to descriptions by Dripps (2003), Dripps and Bradbury (2007), and Steenhuis and Van der Molen (1986).

Model Theory

The SWB code uses a modified Thornthwaite-Mather soil-water accounting method (Thornthwaite and Mather, 1957) to calculate recharge. Recharge is calculated separately for each grid cell in the model domain. Sources and sinks of water within each grid cell are determined on the basis of input climate data and landscape characteristics; recharge is calculated as the difference between the change in soil moisture and these sources and sinks (eq. 1):

$$recharge = (precip + snowmelt + inflow) - \underset{\text{sources}}{} \tag{1}$$
$$(interception + outflow + ET) - \Delta\ soil\ moisture$$
$$\underset{\text{sinks}}{}$$

Each of the water-budget components given in equation 1 is handled by one or more modules within the SWB model. Specific water-balance components are discussed briefly below.

precip—Precipitation data are input as daily values either as a time series at a single gage or as a series of daily Arc ASCII or Surfer grid files created by the user. Precipitation-gage records from an unlimited number of sites may be used if the user supplies precipitation as a series of grid files.

snowmelt—Snow is allowed to accumulate and/or melt on a daily basis. The daily mean, maximum, and minimum air temperatures are used to determine whether precipitation takes the form of rain or snow. Precipitation that falls on a day when the mean temperature minus one-third the difference between the daily high and low temperatures is less than or equal to the freezing point of water is considered to fall as snow (Dripps and Bradbury, 2005).

Snowmelt is based on a temperature-index method. In the SWB code it is assumed that 1.5 mm (0.059 in.) of snow melts (expressed as snow water equivalent) per day per average degree Celsius that the daily maximum temperature is above the freezing point (Dripps and Bradbury, 2005).

inflow—Inflow is calculated by use of a flow-direction grid derived from a digital elevation model to route outflow (surface runoff) to adjacent downslope grid cells. Inflow is considered to be zero if flow routing is turned off.

interception—Interception is treated simply by means of a "bucket" model approach—a user-specified amount of rainfall is assumed to be trapped and used by vegetation and evaporated or transpired from plant surfaces. Daily precipitation values must exceed the specified interception amount before any water is assumed to reach the soil surface. Interception values may be specified for each land-use type and season (growing and dormant).

outflow—Outflow (or surface runoff) from a cell is calculated by use of the U.S. Department of Agriculture, Natural Resources Conservation Service (NRCS) curve number rainfall-runoff relation (Cronshey and others, 1986). This rainfall-runoff relation is based on four basin properties: soil type, land use, surface condition, and antecedent runoff condition.

The curve number method defines runoff in relation to the difference between precipitation and an "initial abstraction" term. Conceptually, this initial abstraction term represents the summation of all processes that might act to reduce runoff, including interception by plants and fallen leaves, depression storage, and infiltration (Woodward and others, 2003). Equation 2 is used to calculate runoff volumes (Woodward and others, 2002):

$$R = \frac{(P - I_a)^2}{(P + [S_{max} - I_a])} \quad P > I_a \tag{2}$$

where

  R is runoff,

  P is daily precipitation,

  S_{max} is maximum soil-moisture holding capacity, and

  I_a is initial abstraction, the amount of precipitation that must fall before any runoff is generated.

The initial abstraction (I_a) term is related to a maximum storage term (S_{max}) as follows:

$$I_a = 0.2S_{max} \qquad (3)$$

The maximum storage term is defined by the curve number for the land-cover type under consideration:

$$S_{max} = \left(\frac{1,000}{CN}\right) - 10 \qquad (4)$$

Curve numbers are adjusted upward or downward depending on how much precipitation has occurred in the previous 5-day period. The amount of precipitation that has fallen in the previous 5-day period is used to describe soil-moisture conditions; three classes of moisture conditions are defined and are called antecedent runoff condition I, II, and III, defined as shown in table 1.

When soils are nearly saturated, as in antecedent runoff condition III, the curve number for a grid cell is adjusted upward from antecedent runoff condition II (eq. 5), to account for generally higher runoff amounts observed when precipitation falls on saturated soil (Mishra and Singh, 2003):

$$CN_{ARC(III)} = \frac{CN_{ARC(II)}}{\left(0.427 + 0.00573 \cdot CN_{ARC(II)}\right)} \qquad (5)$$

Conversely, when soils are dry, as in antecedent runoff condition I, curve numbers are adjusted downward from antecedent runoff condition II (eq. 6) in an attempt to reflect the increased infiltration rates of dry soils (Mishra and Singh, 2003).

$$CN_{ARC(I)} = \frac{CN_{ARC(II)}}{\left(2.281 - 0.01281 \cdot CN_{ARC(II)}\right)} \qquad (6)$$

Between "dry" and "nearly saturated" conditions is antecedent runoff condition II, which represents an average rainfall-runoff relation for moderate soil-moisture conditions.

Curve numbers range from 0 to 100 (Mockus, 1964). If a useful range of curve numbers is defined with a minimum of 30 and maximum of 98, the maximum storage term (S_{max}) varies from a low of about 0.2 in. to a high of about 23 in. Use of an initial abstraction term of $0.2S_{max}$ implies that between 0.04 and 4.6 in. of precipitation must fall before runoff begins; use of an initial abstraction term of $0.05S_{max}$ as suggested by Woodward and others (2003) implies that between 0.01 and 1.15 in. of precipitation must fall before runoff begins.

Frozen ground is tracked by use of a simple continuous frozen-ground index, or CFGI (Molnau and Bissel, 1983):

$$CFGI_i = A \cdot CGFI_{i-1} - T \cdot e^{(-0.4\,K \cdot D)} \geq 0 \qquad (7)$$

where

  $CFGI_i$ is continuous frozen ground index on day i,

  $CFGI_{i-1}$ is continuous frozen ground index on day i–1,

  T is daily mean air temperature (degrees Celsius),

  A is daily decay coefficient,

  K is snow reduction coefficient, and

  D is depth of snow on ground (centimeters).

The values for the coefficients A and K are defined in the same manner as described by Molnau and Bissel (1983): K=0.5 cm^{-1} for above-freezing periods, K=0.08 cm^{-1} for below-freezing periods, and A=0.97. During conditions in which there is no snow cover, the $CFGI$ simply represents the running sum by which the average air temperature deviates from the freezing point of water; snow conditions cause the $CFGI$ to grow or shrink at a slower rate.

The $CFGI$ is applied by allowing for a transition range to be applied through which runoff enhancement ranges from "negligible to strong" (Molnau and Bissel, 1983). A probability of runoff enhancement factor, P_f, is defined as follows:

$$P_f = \frac{CFGI - LL}{UL - LL} \qquad (8)$$

where

  P_f is the probability that runoff will be enhanced by frozen ground conditions,

  $CFGI$ is continuous frozen ground index,

  UL is the upper limit of the $CFGI$, above which frozen ground conditions exist,

  LL is the lower limit of the $CFGI$, below which frozen ground conditions do *not* exist.

Table 1. Definition of antecedent runoff conditions used in the SWB code.

[Precipitation in preceding 5 days, in inches]

Condition	Soil wetness	Dormant season	Growing season
I	Dry	< 0.05	< 1.4
II	Average	0.5–1.1	1.4 – 2.1
III	Near saturation	> 1.1	> 2.1

If no values are assigned, the *CFGI* routine will be ignored. If the *CFGI* option is used, Molnau and Bissel (1983) recommend starting with a value of 83°C-days for the upper limit (*UL*) and a value of 56°C-days for the lower limit (*LL*).

In the SWB code, P_f is used to linearly interpolate between the curve numbers at antecedent runoff condition II and antecedent runoff condition III.

Outflow from a cell becomes inflow to the downslope cell as determined from the flow-direction grid.

evapotranspiration (ET)—The SWB code can use any one of five commonly applied methods to estimate potential evapotranspiration. The methods currently included in the SWB code are:

1. Thornthwaite-Mather (1957),

2. Jensen-Haise (1963),

3. Blaney-Criddle (Blaney and Criddle, 1966; Allen and Pruitt, 1986; Jensen and others, 1990),

4. Turc (1961), and

5. Hargreaves and Samani (1985).

Currently, all methods *except* Hargreaves-Samani (1985) produce an estimate of potential evapotranspiration that is uniform across the model grid. The Hargreaves-Samani method can produce a spatially variable estimate of potential evapotranspiration if supplied with spatially varying minimum and maximum air-temperature grids for each daily timestep.

All methods require specification of daily maximum and minimum air temperatures. The methods other than Thornthwaite-Mather and Hargreaves-Samani require additional data, including data regarding relative humidity, wind speed, and percentage of actual to possible daily sunshine hours.

Δ *soil moisture*—Soil moisture is tabulated by means of the soil-water-balance methods published in Thornthwaite (1948) and Thornthwaite and Mather (1955, 1957). In order to track changes in soil moisture, several intermediary values are calculated, including precipitation minus potential evapotranspiration (*P-PE*), accumulated potential water loss (*APWL*), actual evapotranspiration, soil-moisture surplus, and soil-moisture deficit. These terms are described below.

***P minus PE* (*P – PE*)**—The first step in calculating a new soil moisture value for any given grid cell is to subtract potential evapotranspiration from the daily precipitation (*P – PE*). Negative values of $P - PE$ represent a potential deficiency of water, whereas positive $P - PE$ values represent a potential surplus of water.

***Accumulated Potential Water Loss (APWL)*—**The accumulated potential water loss is calculated as a running sum of the daily $P - PE$ values during periods when $P - PE$ is negative. This running sum represents the total amount of unsatisfied potential evapotranspiration to which the soil has been subjected. Soils typically yield water more easily during the first days in which $P - PE$ is negative. On subsequent days, as the *APWL* grows, soil moisture is less readily given up. The nonlinear relation between soil moisture and the accumulated potential water loss was described by Thornthwaite and Mather (1957) in a series of tables. These tables are incorporated into the SWB code.

Note that the accumulated potential water loss can grow without bound; it represents the cumulative daily *potential* water loss given the *potential* evapotranspiration and observed precipitation.

***Soil moisture, Δ soil moisture*—**The soil-moisture term represents the amount of water held in soil storage for a given grid cell. Soil moisture has an upper bound that corresponds to the soils' maximum water-holding capacity (roughly equivalent to the field capacity); soil moisture has a lower bound that corresponds to the soils' wilting capacity.

When $P - PE$ is positive, the new soil-moisture value is found by adding this $P - PE$ term directly to the preceding soil-moisture value. If the new soil-moisture value is still below the maximum water-holding capacity, the Thornthwaite-Mather soil-moisture tables are consulted to back-calculate a new, reduced accumulated potential water-loss value. If the new soil-moisture value exceeds the maximum water-holding capacity, the soil-moisture value is capped at the value of the maximum water-holding capacity, the excess moisture is converted to recharge, and the accumulated potential water-loss term is reset to zero.

When $P - PE$ is negative, the new soil-moisture term is found by looking up the soil-moisture value associated with the current accumulated potential water-loss value in the Thornthwaite-Mather tables.

***Actual ET*—**When $P - PE$ is positive, the actual evapotranspiration equals the potential evapotranspiration. When $P - PE$ is negative, the actual evapotranspiration is equal only to the amount of water that can be extracted from the soil (Δ *soil moisture*).

***Soil moisture SURPLUS*—**If the soil moisture reaches the maximum soil-moisture capacity, any excess precipitation is added to the daily soil-moisture surplus value. Under most conditions, the soil-moisture surplus value is equivalent to the daily groundwater recharge value.

***Soil moisture DEFICIT*—**The daily soil-moisture deficit is the amount by which the actual evapotranspiration differs from the potential evapotranspiration.

The soil-moisture surplus and deficit terms have no direct bearing on the calculation of recharge; these terms feature rather prominently in the original work by Thornthwaite and Mather (1955, 1957) and are included here for completeness.

Data Requirements for Application of the SWB Model

The SWB model requires the user to provide tabular climatological and gridded land-surface data in order to calculate a water budget and a recharge estimate for each grid cell (table 2).

Table 2. Data requirements for application of the SWB model.

Data type
Gridded (ARC ASCII or Surfer grid)
Land use/land cover
Flow direction (D8)
Hydrologic soil group
Available water capacity
Tabular
Soil and land use properties lookup table
Climate at a single station
Matrix of soil-water retention for given accumulated potential water loss

Four gridded datasets are required: (1) hydrologic soil group, (2) land-use/land-cover, (3) available soil-water capacity, and (4) surface-water flow direction.

In addition to the gridded land-surface data, the model requires tabular daily climatological data. At a minimum, the model needs daily precipitation (in inches), daily average air temperature (in degrees Fahrenheit), daily maximum air temperature (in degrees Fahrenheit), and daily minimum air temperature (in degrees Fahrenheit), but it may require additional data depending on the formulation of the evapotranspiration equation the user specifies for the water-budget calculations. Additional optional data types include: (1) daily average wind speed (in meters per second), (2) daily average relative humidity (in percent), (3) daily maximum relative humidity (in percent), and (4) daily percentage of possible sunshine (in percent).

Finally, a lookup table must be supplied in order to assign curve numbers, interception values, rooting depths, and maximum daily recharge values to each combination of hydrologic soil group and land-use/land-cover type.

The relation between each of these data types and the SWB code is shown in figure 1. Further details on the input data formats and requirements may be found in the subsection "Use of the SWB Model" below.

Summary of Major Differences Between the Original and Current SWB Codes

The original Visual Basic for Applications (VBA) code of Dripps (2003) and Dripps and Bradbury (2007) differs in several ways from the current Fortran SWB model code. In most cases, the differences do not change the pattern or quantity of calculated recharge appreciably; other changes are expected to result in local differences in calculated recharge. Major differences between the Visual Basic code and the Fortran SWB code include the following:

1. **Soil-moisture determination.** In the VBA code, polynomials were used to approximate Thornthwaite-Mather soil-moisture tables, rather than the tables themselves. Transformations between soil-moisture values to accumulated potential water-loss values were not entirely mass conservative. The SWB code uses interpolated values from the original Thornthwaite-Mather tables.

2. **Integer arithmetic.** In the VBA code integer arithmetic and rounded or truncated values were used in the calculation of many components of the soil-water balance, including the curve number, maximum soil-water capacity, and average daily temperature. The SWB code uses real values in all calculations.

3. **Date calculations.** The SWB code now requires the climatological data file to refer to each simulation day by its actual month, day, and year of observation. The code accounts for leap years; data from February 29 in a leap year need not (and should not) be excluded from the climate data file. The original VBA code did not allow for inclusion of an extra day during a leap year.

4. **Flow directions.** If flow directions from two adjoining cells face each other, the original VBA model made only the first cell a closed depression. By contrast, the Fortran code defines both cells as depressions whenever two neighboring cells have flows in opposing directions. This effectively spreads the same volume of recharge out over a larger land-surface area.

5. **Rain on snow.** In the original VBA code, 50 percent of precipitation falling as rain when snow is present on the ground was assumed to run off immediately and was modeled by reducing the incoming precipitation by 50 percent. In the SWB code, rainfall that falls on snow is converted to runoff by use of frozen ground properties (described in the next section) and is variable in time and space depending on soil conditions and properties.

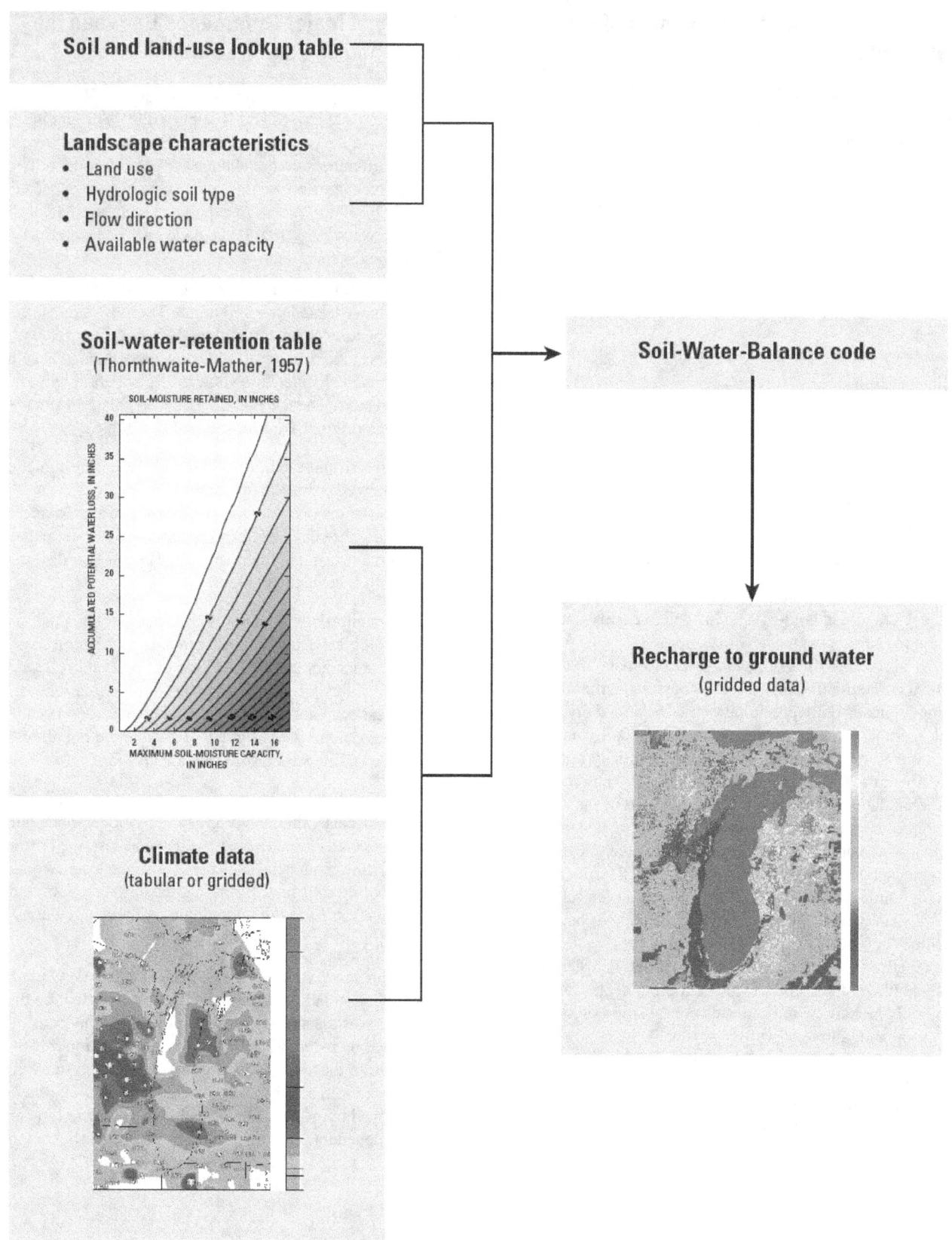

Figure 1. Interaction between Soil-Water-Balance code and data.

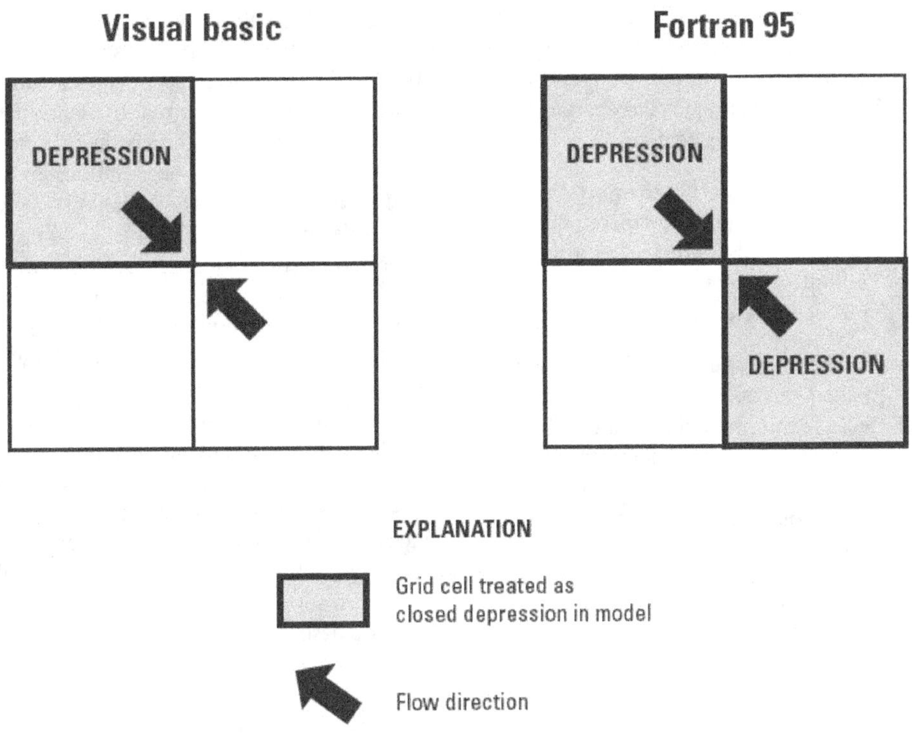

Visual basic

Fortran 95

EXPLANATION

Grid cell treated as
closed depression in model

Flow direction

Figure 2.   Definition of closed depression based on flow-direction-grid inputs.

6. **Urban-area recharge.** In the original VBA code, it is assumed that the calculated recharge beneath an impervious surface is further reduced by some fraction. The VBA code accounts for impervious surfaces by reducing the calculated daily recharge estimates by the percent imperviousness (a fixed-value factor specified in the code, ranging from 0.05 to 0.88). The SWB model does not perform this correction; rather, it is assumed that the specification of an SCS curve number for urban land uses already reflects the impact of significant impervious areas.

7. **Soil-water capacity.** Similar to soil-moisture determinations described above, maximum soil-water capacity calculations in the original VBA code involved approximation of root-zone depths by means of polynomial expressions. The SWB code requires that the rooting-depth be specified for each combination of land use and soil type.

8. **Treatment of open-water cells.** Recharge in cells is not calculated where (a) the soil-water capacity is near zero, or (b) the land-cover type of the cell has been explicitly identified as open water via the OPEN_WATER_LAND_USE directive within the input control file. Open-water cells are assumed to drain through surface-water features, which are not explicitly considered in the SWB code.

Extensions and Additional Capabilities of the SWB Code

In addition to the differences described in the previous section, the capabilities of the current version of the SWB code have been extended beyond those of the original VBA code. The added capabilities include the following:

1. **Multiple-year simulation capability.** The SWB code may be run for multiple years in a single program execution by specifying additional single-station climate data files.

2. **Continuous frozen ground index (CFGI).** For each grid cell, a CFGI is calculated on the basis of daily air temperatures (Molnau and Bissell, 1983). The user may specify both a value above which the ground should be considered frozen and a value below which the ground is considered unfrozen, both in degree-Celsius-days. When the CFGI lies between these user-specified values, the code shifts the runoff curve numbers from antecedent runoff condition II toward antecedent runoff condition III, which has the effect of increasing the individual curve numbers as well as runoff from the grid cell.

3. **Maximum recharge rate.** A maximum recharge rate may be specified for each combination of land cover and hydrologic soil group. The user might specify this to be consistent with the saturated vertical conductivity used in

the underlying MODFLOW model. If this option is used, recharge in excess of the maximum specified recharge rate is converted to "rejected recharge" and is removed from the model grid. It is assumed that rejected recharge leaves the model domain via surface-water features.

4. **Gridded precipitation and temperature input.** Precipitation and minimum and maximum air temperatures may be specified as a series of Arc ASCII or Surfer grid files. This allows for an unlimited number of precipitation and temperature stations to be included in a simulation.

5. **Hargreaves-Samani (1985) evapotranspiration process module.** This ET method was added for use with gridded climate data. Because this method estimates ET on the basis of maximum and minimum air temperatures, a more site-specific ET calculation may be made without requiring the additional input of gridded solar radiation, relative-humidity, or wind-speed data.

6. **Alternate initial abstraction definition.** The user may choose either the original curve number methodology, in which the initial abstraction term is defined as equal to $0.2S_{max}$, or may use an alternate definition in which this term is defined as equal to $0.05S_{max}$ (Woodward and others, 2003). The use of a smaller initial abstraction term may be more appropriate for continuous simulations, and it will increase runoff from smaller precipitation events (Woodward and others, 2003).

Model Limitations and Assumptions

The original concept behind the SWB code was to allow for the spatial distribution of groundwater recharge to be quickly and easily calculated on the basis of readily available data and a standardized set of parameters (Dripps, 2003). The version of the SWB model described here retains most of the features that made the original model attractive from the standpoint of practical application (Dripps, 2003; Dripps and Bradbury, 2007). Despite the possible limitations given below, the SWB model approach should be capable of generating reasonable annual or monthly mean groundwater recharge estimates at the scale of a small catchment (Dripps and Bradbury, 2007). In order to do so, however, the user will have to upscale the daily results offered by the SWB model and average or filter the results over a larger area. The relative spatial variability and pattern of recharge between catchments should also be of great value, particularly if these estimates can be corroborated with recharge estimates generated from streamgages or observation wells. Comparing SWB-calculated recharge estimates to those estimated from streamflow records, or from a groundwater model calibrated to stream fluxes is recommended.

As with any numerical model, the burden is on the user to preprocess the input grids in the most appropriate manner. If the user has not done this, then SWB will generally halt after giving the user a description of the problem it detects for an input grid. Although the SWB code can certainly be applied using only available data and a standard set of curve numbers, it would be prudent to treat the results with caution, as one should with any model output. In addition, certain underlying theoretical limitations should be kept in mind when interpreting SWB model output. These limitations are discussed below.

Runoff routing—The inclusion of overland-flow routing in the code ensures that runoff from an upslope grid cell has one or more opportunities to contribute to infiltration in the cells that are downslope from it. However, all runoff from a cell is assumed to infiltrate in downslope cells or be routed out of the model domain on the same day in which it originated as rainfall or snowmelt.

In addition, once water is routed to a closed surface depression and evapotranspiration and soil-moisture demands are met, the only loss mechanism is recharge. This results in cases where maximum recharge values of hundreds or thousands of inches per year are calculated.

These extremely high values are unrealistic and likely result from surface storage of water not being accounted for. The code described here allows the user to enter a maximum recharge rate for each land-cover and soil-group combination. This feature offers a way to restrict the estimated recharge values to a more reasonable range; however, the rejected recharge is nonetheless removed from the model domain on the same day in which it originated as precipitation or snowmelt.

Groundwater/surface-water interaction—Interactions between surface-water and groundwater features are not simulated in the SWB code and could not be without significantly increasing the complexity of the model. In locations where the water table is beneath the bottom of the root-zone, the SWB code should be capable of producing reasonable annual or monthly values. The depth from the bottom of the root zone to the top of the water table is not considered in the estimation of recharge; there may be a significant time of travel through the unsaturated zone. Coupling the SWB code with an unsaturated-zone code that could route water to the water table, such as the MODFLOW UZF Package (Niswonger and others 2006), would be one way to address this limitation.

In areas with wetlands, springs, lakes, or other landscape features where the water table is close to the land surface, the SWB code can be expected to perform poorly; there is currently no provision for recharge rejection via saturation excess other than by specifying a maximum recharge rate for a particular combination of land use and soil type.

Curve number method—In the current version of the SWB model, it is assumed that infiltration is the sum of net precipitation, snowmelt, and inflow, minus the runoff calculated by means of the NRCS curve number method. Runoff calculation at a plot or field scale in a continuous simulation by means of the curve number method may be beyond the limits of the method. The list of perceived limitations associated with the curve number method includes the following (Garen and Moore, 2005):

- method cannot be used to identify runoff processes, source areas, or flow paths;

- method is a watershed-scale method that should not be applied at a plot or field (or grid cell) scale; and

- method was developed to evaluate floods and was not designed to simulate daily flows of ordinary magnitude.

In addition, it has been suggested that the curve number itself is not constant but varies from event to event and that the antecedent-runoff condition explains only some of this variability (Hjelmfelt, 1991).

Given variability in the curve numbers themselves, as well as the other limitations of the curve number method, it is reasonable to treat the standard curve number table values merely as starting points; ideally, the curve numbers should be verified by use of observed paired precipitation-runoff data (Hawkins, 1993).

The SWB code contains an alternative method for calculating runoff that incorporates a much smaller initial abstraction term. Use of this alternative method for calculating the initial abstraction may be more appropriate for continuous simulation (Woodward and others, 2003). Users of the SWB code have the option of defining the initial abstraction term as $I_a = 0.05S_{max}$, compared to $I_a = 0.2S_{max}$. The use of this smaller initial abstraction term results in more runoff generation for areas with low curve numbers and for storms of smaller magnitude. If the smaller initial abstraction term is used, curve numbers are automatically scaled by the SWB code to maintain an appropriately shaped rainfall-runoff curve (Woodward and others, 2003).

The modular design of the SWB code makes it feasible to add new process modules relatively easily. Although there are no immediate plans to do so, future versions of SWB could include an implementation of the Green-Ampt infiltration method (Green and Ampt, 1911), and an enhanced ability to route and store overland flow.

Snowmelt and infiltration—For temperate areas that experience snowfall and snowmelt, the SWB model is sensitive to snowmelt, and in particular, to how snowmelt translates into surface runoff. The addition of a continuous frozen ground index (CFGI) to the SWB code offers a simple way to approximate the effects of frozen ground. Spring runoff may be increased by lowering the setpoints at which the ground is considered to change from unfrozen to frozen; lowering the CFGI setpoints has the effect of increasing the amount of time that the runoff curve numbers are shifted toward antecedent runoff condition III.

Other modelers have altered the curve number in an attempt to simulate runoff from frozen ground. For example, Carroll and others (2005) assigned a separate set of curve numbers to soils considered to be frozen and another set of curve numbers to soils considered to be unfrozen. Despite this, there is no theoretical basis supporting the derivation of a frozen-ground curve number, so its use in the SWB code is

primarily for expediency and consistency with other model input considerations.

Because the CFGI is based on air temperatures, the SWB code is unable to resolve differences in snowmelt timing between grid cells with differing ground-surface orientation relative to the sun (aspect).

Climate variability—Year-to-year climate variability causes corresponding variability in calculated recharge values. Use of multiple years of climate data should help to minimize the effect of year-to-year climate variability on estimated recharge values.

Use of the SWB Model

Basic application of the SWB code is relatively straightforward. This section describes the requirements for application of the SWB code, including the directory structure and input files, the types of output available, and the various control-file options.

Required Directory Structure

The code is set up with certain expectations about how the project directory will be structured. If the code is run and the expected directory structure is not found, program execution will halt and an error message will be given. The user should ensure that this directory structure exists before running the SWB code.

It is desirable to begin SWB program execution from within the top-level directory because the code uses relative pathnames to keep track of input and output. The required directory structure is shown in figure 3.

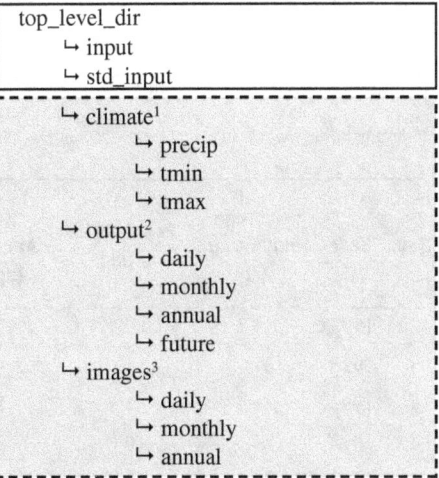

[1]Required only if gridded climate data is used.
[2]Required only if gridded output is requested.
[3]Required only if image output is requested.

Figure 3. Required directory structure.

If gridded precipitation or temperature files are used, an additional subdirectory (or subdirectories) directly beneath the top-level directory may be created to hold the additional files. Figure 3 includes an example of how this might be done. The associated statements within the control file must reflect the user's directory structure. If, for example, the directory structure shown in fig. 3 was used in a project along with daily gridded precipitation values, the associated control-file statement would look like the one shown below; note that climate\precip\ in the example specifies subdirectory name relative to the name of the top-level directory.

Example control-file statement:

```
PRECIPITATION Arc_GRID climate\precip\file_prefix
```

More details regarding the use of gridded precipitation and temperature data are included in the sections that follow.

Input Files

This section describes the format and content of the various tabular and gridded data files required by the SWB code.

Tabular Files

At a minimum, a single file containing climate data is required to use the SWB model. The SWB code contains routines to convert between the Gregorian date and the Julian day number, and accounts for leap years automatically. The SWB code ignores any lines in the file that begin with a pound sign (#). Unused data items in the table may be set to any value but must be present. An example of the required format for this file is given in table 3; unused data items in table 3 have been set to -99999. Note that there should not be any missing values in the fields that are actively being used in the simulation.

Minimum, maximum, and average daily temperature data are required inputs, as is the daily precipitation value. Additional (optional) data values are needed if an evapotranspiration calculation method other than the Thornthwaite-Mather or Hargreaves-Samani method is desired. The additional data types include the following:

- Percentage of possible sunshine: the ratio of actual hours of sunshine relative to the total possible daily hours of sunshine, in percent.

- Minimum relative humidity: the daily minimum recorded relative humidity, in percent.

- Average relative humidity: the daily average relative humidity, in percent.

- Wind speed: the daily average wind speed, expressed in meters per second and measured at 2 m above land surface.

Table 4 lists the additional climate data requirements for each available evapotranspiration (ET) calculation method included in the SWB code. Note that at this time the only ET method suitable for use with gridded precipitation data is the Hargreaves-Samani method; all other methods require additional gridded data, such as relative humidity and wind speed, in order to properly apply them in a distributed fashion.

Gridded Files

The SWB code requires gridded data for four data types: (1) hydrologic soil group, (2) land-use/land-cover, (3) available soil-water capacity, and (4) surface-water flow direction. **For the code to work, all gridded data sets must share a common datum and projection, grid-cell size, and grid extent.** The SWB code has no capability to change or convert

Table 3. Example of the data elements contained in the tabular climate data file.

[F, degrees Fahrenheit; in., inches; m/s, meters per second; %, percent]

Month	Day	Year	Average temperature (°F)	Precipitation (in.)	Relative humidity (%) [optional]	Maximum temperature (°F)	Minimum temperature (°F)	Wind speed (m/s) [optional]	Minimum relative humidity (%) [optional]	Possible % sunshine [optional]
1	2	1993	21.5	0.09	79.5	33	10	5.41	67	0
1	3	1993	37	.1	94	41	33	4.87	92	0
1	4	1993	31.5	.17	88.5	39	24	4.6	81	0
1	5	1993	23.5	.003	73	28	19	3.17	61	40
1	6	1993	20.5	.003	68	26	15	3.08	55	49
1	7	1993	16	.003	77.5	22	10	2.86	71	0
1	8	1993	14	.003	70.5	22	6	3.93	57	64
1	9	1993	18.5	.003	76	21	16	7.24	71	0

< Data file continues with a separate line for each day of the year. >

projections and datums; the program will terminate with an error message if the grids do not share common grid-cell sizes and extents.

ArcInfo, ArcMap, ArcView, or Surfer may used to code and generate the gridded ASCII input files. An example of the first 7 lines of an Arc ASCII Grid file are shown in figure 4.

Land Use : Integer Grid

The model uses land-use/land-cover information, together with the available soil-water capacity, to calculate surface run-off and assign a maximum soil-moisture holding capacity for each grid cell. This version of the model can handle any arbitrary land-use classification method as long as the accompanying land-use lookup table contains curve-number, interception, maximum-recharge, and rooting-depth data for each land-use type contained in the grid.

The original model required that land-use classifications follow a modified Anderson Level II Land Cover Classification (Anderson and others, 1976). The modified Anderson Level II classification scheme of Dripps (2003) is given in table 5.

Hydrologic Soil Group : Integer Grid

The U.S. Department of Agriruglture, Natural Resources Conservation Service (NRCS), formerly the Soil Conservation Service (SCS), has categorized more than 14,000 soil series within the United States into one of four hydrologic soil groups (A – D) on the basis of infiltration capacity. NRCS hydrologic soil group information may be input to the model as an Arc ASCII or Surfer integer grid with values ranging from 1 (soil group A) to 4 (soil group D). The NRCS hydrologic soil group A soils have a high infiltration capacity and, consequently, a low overland flow potential. Group D soils, in contrast, have a very low infiltration capacity and, consequently, a high overland flow potential (table 6).

```
ncols          356
nrows          331
xllcorner      527499.78835059
yllcorner      274275.62601013
cellsize       100
NODATA_value   -99999
41 41 41 41 41 41 41 41 41 31 31 31 31 31 31 41 41 41 41 41 41 41
...
```

Figure 4.   Example of Arc ASCII grid input-file format.

Table 4.   Data requirements for various potential evapotranspiration methods.

[F, degrees Fahrenheit; %, percent; m/s, meters per second]

Method	Mean air temperature (°F)	Minimum air temperature (°F)	Maximum air temperature (°F)	Mean relative humidity (%)	Wind velocity (m/s)	Minimum relative humidity (%)	Percent possible sunshine (%)	Suitable for use with gridded precip and air temperature data
Thornthwaite–Mather	✓							
Jensen-Haise	✓						✓	
Blaney-Criddle (FAO BC)	✓	✓	✓		✓	✓	✓	
Turc	✓			✓			✓	
Hargreaves-Samani	✓	✓	✓					✓

Table 5.   Modified Anderson Level II land-use classification scheme.

Level I Classification	Modified Level II Classification (Dripps, 2003)
1 - Urban or built up	
	11 - Residential
	12 - Commercial services
	13 - Industrial
	14 - Transportation
	17 - Golf course/park/open space
2 - Agricultural land	
	22 - Orchards
	25 - Shallow-rooted crops (such as spinach, peas, beets, carrots)
	26 - Moderately-Deep rooted crops (such as corn, cotton, tobacco, cereal grains)
	27 - Deep-rooted crops (alfalfa)
	28 - Fallow
3 - Pasture/rangeland	
	31 - Pasture/rangeland
4 - Forest land	
	41 - All forest types—deciduous, evergreen, and mixed forest
5 - Water	
	51 - All water bodies—streams, canals, lakes, reservoirs, bays, estuaries
6 - Wetland	
	61 - Forested wetland
	62 - Nonforested wetland
7 - Barren land	
	72 - Beaches/sandy areas
	74 - Bare exposed rock

Table 6.   Infiltration rates for Natural Resources Conservation Service hydrologic soil groups (Cronshey and others, 1986).

Soil Group	Infiltration rate
A	> 0.3 inch per hour
B	0.15–0.3 inch per hour
C	0.05–0.15 inch per hour
D	< 0.05 inch per hour

In most straightforward applications of the SWB code, the hydrologic soil groups (A through D) would be converted to a numerical code (1 through 4), and exported as an Arc ASCII or Surfer grid. The land-use lookup table could then be populated with values pulled directly from the NRCS tables applicable to each land-use/land-cover type.

The SWB code will support any arbitrary number of soil groups that the user wishes to define. The user must specify a curve number, maximum infiltration rate, and rooting depth for each combination of land use and soil type.

Surface-Water Flow Direction : Integer Grid

The SWB code requires a flow-direction grid for the entire model domain. The code uses this grid to determine how to route overland flow between cells. The user must create the flow direction grid consistent with the D8 flow-routing algorithm (O'Callaghan and Mark, 1984), with flow directions defined as shown in figure 5. The original D8 algorithm assigns a unique flow direction to each grid cell by finding the steepest slope between the central cell and its eight neighboring cells.

Some implementations of the D8 algorithm are capable of accommodating flows to several neighboring cells by assigning a combination of the numbers shown in figure 5 (Jenson and Domingue, 1988). For example, consider a cell (blue cell in fig. 5) that has the same downhill slope in the direction of the two neighboring cells to the west and northwest. Flow could reasonably be expected to go to both of these neighboring cells. Flow only to the cell to the west would be assigned a flow direction of 16; flow only to the cell to the northwest would be assigned a flow direction of 32. Flow to both cells would be indicated by adding the two individual flow-direction values, resulting in a flow-direction value of 48. **Note that the SWB model is not designed to accommodate flows to more than one cell.**

In the SWB code, a cell for which the flow-direction value is not a power of 2 (as shown in fig. 5) is considered to indicate a closed depression. The SWB code does not attempt to split flows between two or more cells; when a cell with more than one possible flow direction is encountered, it is identified as a closed depression. The SWB code allows no further surface runoff to be generated or ponding to occur but instead requires water in excess of the soil-moisture capacity to contribute to recharge.

ArcInfo software, with the **GRID** extension, may be used to generate a D8 flow-direction grid from a digital elevation model (DEM) file using the **GRID** command `FLOWDIRECTION`.

Available Soil-Water Capacity : Real Number Grid

The SWB model uses soil information, together with land-cover information, to calculate a maximum soil water-holding capacity for each grid cell. The maximum soil-water capacity is calculated as

$$\textit{maximum soil water capacity} = \atop \textit{available soil water capacity} \cdot \textit{root-zone depth} \qquad (7)$$

Soil surveys, which include an estimate of the available water capacity or textural information, are typically available through the state offices of the NRCS, or on the world-wide web at http://soils.usda.gov/. Each soil type or soil series within the model area must be assigned an available water capacity. If data for available water capacity are not available,

Figure 5. Numerical definition of D8 flow directions.

32	64	128
16	Central cell (flow origin)	1
8	4	2

Table 7. Estimated available water capacities for various soil-texture groups.

Soil texture	Available water capacity (inches per foot of thickness)
Sand	1.20
Loamy sand	1.40
Sandy loam	1.60
Fine sandy loam	1.80
Very fine sandy loam	2.00
Loam	2.20
Silt loam	2.40
Silt	2.55
Sandy clay loam	2.70
Silty clay loam	2.85
Clay loam	3.00
Sandy clay	3.20
Silty clay	3.40
Clay	3.60

the user can use soil texture to assign a value, shown in table 7 (Dripps, 2003; original source table 10, Thornthwaite and Mather, 1957).

The available water capacity of a soil is typically given as inches of water holding capacity per foot of soil thickness. For example, if a soil type has an available water capacity of 2 in/ft and the root-zone depth of the cell under consideration is 2.5 ft, the maximum water capacity of that grid cell would be 5.0 in. This is the maximum amount of soil-water storage that can take place in the grid cell. Water added to the soil column in excess of this value will become recharge.

Note that a grid containing the maximum soil-water capacity may be input directly into the SWB code, bypassing the internal calculation of the maximum soil-water capacity.

Lookup Tables

The SWB code uses two lookup tables to calculate model parameters on the basis of grid-cell properties. The first of these files is the land-use lookup table, which contains NRCS curve number, rooting depth, interception, and maximum daily recharge information specific to each land-use type in the model application. The land-use lookup table supplied with this version of the code contains Anderson Level II land-use classifications described by Dripps (2003). The land-use lookup table may be added to or modified to allow the model to work with any arbitrary method of land-use classification; data elements and an example annotated lookup table are given in tables 8 and 9, respectively.

The second standard file contains an extended version of the Thornthwaite-Mather soil-water-retention tables, which relate the accumulated potential water loss to the amount of soil moisture retained over a range of soil-water capacities. The soil-water-retention table should not need user modification.

Land-Use Lookup Table

The land-use lookup table allows the user to specify curve numbers, maximum recharge rates, and root-zone depths for each soil type within a given land-use type. In addition, a precipitation interception amount may be specified for each land-use type.

The first line of this file must begin with NUM_LANDUSE_TYPES ##, where ## is the number of active land-use types contained in the table. The second line of the file must contain the text NUM_SOIL_TYPES ##, where ## represents the number of distinct soil types represented within the table. The remainder of the file is a tab-delimited text file having one line for each land use specified within the land-use grid. Any line that begins with a # will be ignored by the SWB code. Data items are defined as listed in table 8.

An example of a land-use lookup table is given in table 9. The table has been formatted for ease of reading; if the lines, colors, and text-wrapping are removed, table 9 works perfectly with the SWB model. The curve-number, maximum-recharge, and root-zone-depth blocks are always defined in order of ascending soil-type number.

The root-zone depths in the SWB model are one of the more important parameter sets. Thornthwaite and Mather (1957) note that "one factor which complicates the relation between depth of rooting of a plant and the type of vegetation is that the same plants will send roots to different depths in different types of soil. Thus in a sandy soil plants tend to be more deeply rooted than in silts and clays."

Table 8. Data elements contained in the land-use lookup table.

Column number	Description	Notes
1	Land-use code	Integer value corresponding to the integer values contained in the land-use ARC ASCII grid.
2	Land-use description	Not used by model; for use by user to document the description of the land-use corresponding to the integer land-use code.
3	Assumed impervious area	Not used by model; for use by user to document assumed impervious area associated with the land-use code.
4–7*	NRCS base curve numbers	NRCS base curve numbers for hydrologic soil groups A–D, respectively. The curve numbers are those associated with antecedent runoff condition II. A curve number must be specified for each soil type.
8–11*	Maximum infiltration rates	Maximum infiltration rates (inches/day) for each soil type.
12,13	Interception storage values	Interception storage values for growing season and dorment season.
14–17*	Depth of root zone	Root–zone depth, in feet, for each soil group A–D.
18,19	Reference	Not used by model; for use by users in documenting the sources of information placed into the table.

*Column numbering shown reflects specification of only four soil types. If more than four soil types are present, the column numbering will shift accordingly.

Table 9. Annotated example land-use lookup table.

LU code	Description	Assumed imperviousness	CURVE NUMBERS				
			Soil #1	Soil #2	Soil #3	Soil #4	Soil #5
			Clay till (C)	Loamy till (B)	Fine (D)	MED/CRS (A)	Organic (B/C)
11	Open water	Not applicable	100	100	100	100	100
12	Perennial ice/snow	Not applicable	40	40	40	40	40
21	Low density residential	30	83.4	80	85.6	74.6	81.7
22	High density residential	65	89.6	88	90.2	86	88.8
23	Commercial/industrial transportation	85	93.5	93	93	93.1	93.3
31	Bare exposed rock / sand / clay (assume similar to dirt road)	Not applicable	88.1	82	90.5	71.1	85.1
32	Quarries/gravel pits (assume similar to comercial)	0	94.7	92	95.8	87.2	93.4
33	Transitional (assume similar to newly gradedareas)	0	90.8	86	92.6	77.5	88.4
41	Deciduous forest	0	60	50	68.5	32	55
42	Evergreen forest	0	63.9	55	71.4	39.1	59.5
43	Mixed forest	0	67.8	60	74.2	46.2	63.9
51	Shrubland (assumed same as parkland)	0	74.8	69	79.3	59	71.9
61	Orchard	0	71.7	65	77.1	53.3	68.4
71	Grasslands / herbaceous (deep-rooted ag)	0	80.3	76	83.3	68.9	78.1
81	Pasture (assumed type pasture, good condition)	0	88.1	86	89	83.1	87
82	Row crops (shallow-rooted agriculture)	0	88.9	87	89.6	84.5	87.9
83	Small grains (moderate-rooted agriculture)	0	81.8	78	84.5	71.8	79.9
84	Fallow (assumed type fallow, bare soil)	0	88.9	87	89.6	84.5	87.9
85	Urban/recreational grasses (assumed type open space, fair)	0	79.5	75	82.8	67.5	77.3
91	Forested wetland	0	90.4	89	90.7	87.4	89.7
92	Wetland	0	91.2	90	91.3	88.8	90.6

Table 9. Annotated example land-use lookup table—Continued.

| | MAX RECHARGE (INCHES PER DAY) | | | | | INTERCEPTION | |
| | Soil #1 | Soil #2 | Soil #3 | Soil #4 | Soil #5 | | |
LU code	Clay till (C)	Loamy till (B)	Fine (D)	MED/CRS (A)	Organic(B/C)	Growing season	Non-growing season
11	0.12	0.6	0.24	2	0.6	0	0
12	.12	.6	.24	2	.6	0	0
21	.12	.6	.24	2	.6	.0835	0
22	.12	.6	.24	2	.6	.0835	0
23	.12	.6	.24	2	.6	.0625	0
31	.12	.6	.24	2	.6	0	0
32	.12	.6	.24	2	.6	0	0
33	.12	.6	.24	2	.6	.09	0
41	.12	.6	.24	2	.6	.05	0
42	.12	.6	.24	2	.6	.05	0
43	.12	.6	.24	2	.6	.05	0
51	.12	.6	.24	2	.6	.0625	0
61	.12	.6	.24	2	.6	.05	0
71	.12	.6	.24	2	.6	.09	0
81	.12	.6	.24	2	.6	.09	0
82	.12	.6	.24	2	.6	.09	0
83	.12	.6	.24	2	.6	.09	0
84	.12	.6	.24	2	.6	0	0
85	.12	.6	.24	2	.6	.0625	0
91	.12	.6	.24	2	.6	.05	0
92	.12	.6	.24	2	.6	0	0

The fact that plants will send roots to different depths in different types of soil is the motivation behind table 10. The SWB code requires that a root-zone depth be entered explicitly for each land-use and soil-type combination. The values in table 10 are an excellent place to start. However, Thornthwaite and Mather's work was motivated by a need to estimate the surplus and deficit of soil water for irrigation needs, and may not necessarily represent ideal values for the purposes of groundwater-recharge estimation.

Also note that the water-holding capacities shown in table 10 were developed as part of the soil-water-balance methodology of Thornthwaite and Mather (1957). However, in a comparison of potential evaporation functions, Vörösmarty and others (1998) show that the Thornthwaite-Mather evapotranspiration calculation method tends to be biased low (as much as -94 mm/yr) relative to other common methods; use of the table 10 water-holding capacities with other evapotranspiration methods may result in overestimation of the amount of evapotranspiration and underestimation of recharge.

The land-use file must be tab delimited. One way to create a tab-delimited file is to edit the file in Microsoft Excel and select the **File, Save As,** and **Text (Tab Delimited)(*.txt)** menu items.

Soil-Moisture Retention Table

The soil-moisture-retention table is used to calculate changes in soil moisture during periods of unsatisfied potential evapotranspiration. The code uses the accumulated potential water loss along with the maximum soil-moisture capacity of a grid cell to determine the amount of soil moisture that would remain under such conditions. The table included with the SWB code is a modified version of the original tables by Thornthwaite and Mather (1957).

The original Thornthwaite-Mather tables contained values for maximum soil-moisture capacities ranging from 1.0 to 16.0 in. The modified table extrapolates values with maximum soil-moisture capacities below 1.0 in. and above 16.0 in.,

Table 9.   Annotated example land-use lookup table—Continued.

| LU code | ROOT ZONE DEPTH (FEET) | | | | |
| | Soil #1 | Soil #2 | Soil #3 | Soil #4 | Soil #5 |
	Clay till (C)	Loamy till (B)	Fine (D)	MED/CRS (A)	Organic (B/C)
11	0	0	0	0	0
12	0	0	0	0	0
21	2	2	2	2	2
22	2	2	2	2	2
23	2	2	2	2	2
31	1	1	1	1	1
32	1	1	1	1	1
33	1	1.81	1.39	1.67	1.52
41	1.74	1.97	1.82	2	1.83
42	1.74	1.97	1.82	2	1.83
43	2.17	2.79	2.61	2.67	2.69
51	2.59	3.61	3.4	3.33	3.54
61	2.59	5.37	3.47	5.37	3.75
71	2.11	3.61	3.4	3.33	3.54
81	2.11	3.61	3.4	3.33	3.54
82	.63	2	1.67	1.67	1.76
83	2	3.33	2.73	3.05	2.84
84	.5	.5	.5	.5	.5
85	2.59	3.61	3.4	3.33	3.54
91	4.5	4.5	4.5	4.5	4.5
92	4.5	4.5	4.5	4.5	4.5

resulting in a table that spans maximum soil-moisture capacities from 0.5 to 17.5 in.

In addition, table values were extrapolated to cover a maximum accumulated potential water loss of as much as 40.7 in. The original Thornthwaite-Mather tables stopped once the remaining soil moisture for a given maximum soil-moisture capacity approached about 1.0 in. Discontinuities in the table values caused instabilities in the SWB code because of the nature of the algorithm used to look up remaining soil-moisture values. Therefore, the tables were extrapolated to yield accumulated potential water-loss values that cover the range down to 40.7 in.

The relation between accumulated potential water loss and remaining soil moisture, as implemented in the SWB model, is shown in figure 6.

Output Files

The SWB code can supply many different types of output at a daily, monthly, or yearly frequency. Output types include tabular, gridded, and image data files. The specific model output types are described in the following sections.

Tabular Files

The SWB code produces four text files summarizing the model run. The files are written in the same subdirectory in which the SWB executable resides. Three of these files (statistics files) are overwritten each time the model is run. One of these files (the log file) is created anew with each execution of the model.

Table 10. Provisional water-holding capacities with different combinations of soil and vegetation (Thornthwaite and Mather, 1957).

Soil type	Available water		Root zone	
	millimeters per meter	inches per foot	meters	feet
Shallow-rooted crops (spinach, peas, beans, beets, carrots, etc.)				
Fine sand	100	1.2	0.50	1.67
Fine sandy loam	150	1.8	.50	1.67
Silt loam	200	2.4	.62	2.08
Clay loam	250	3.0	.40	1.33
Clay	300	3.6	.25	0.83
Moderately deep-rooted crops (corn, cotton, tobacco, cereal grains)				
Fine sand	100	1.2	.75	2.05
Fine sandy loam	150	1.8	1.00	3.33
Silt loam	200	2.4	1.00	3.33
Clay loam	250	3.0	.80	2.67
Clay	300	3.6	.50	1.67
Deep-rooted crops (alfalfa, pastures, shrubs)				
Fine sand	100	1.2	1.00	3.33
Fine sandy loam	150	1.8	1.00	3.33
Silt loam	200	2.4	1.25	4.17
Clay loam	250	3.0	1.00	3.33
Clay	300	3.6	.67	2.22
Orchards				
Fine sand	100	1.2	1.50	5.00
Fine sandy loam	150	1.8	1.67	5.55
Silt loam	200	2.4	1.5	5.00
Clay loam	250	3.0	1.00	3.33
Clay	300	3.6	.67	2.22
Closed mature forest				
Fine sand	100	1.2	2.50	8.33
Fine sandy loam	150	1.8	2.00	6.66
Silt loam	200	2.4	2.00	6.66
Clay loam	250	3.0	1.60	5.33
Clay	300	3.6	1.17	3.90

SOIL MOISTURE RETAINED, IN INCHES

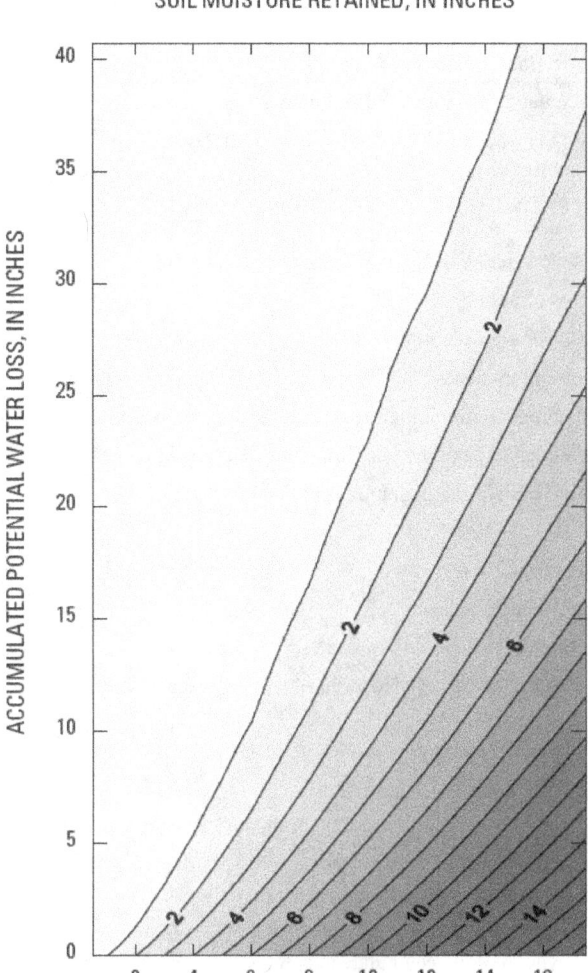

Figure 6. Soil-moisture-retention table (based on Thornthwaite and Mather, 1957).

recharge_daily_statistics—This file contains minimum, mean, and maximum values for all major components of the water balance for all grid cells in the model domain on a given day. The units are inches, except for the continuous frozen ground index (CFGI), which is in degree-Celsius-days.

recharge_daily_report—This is similar to the recharge_daily_statistics file, except that the units are expressed in acre-feet of water. The reported values represent the individual daily values summed over the entire model domain multiplied by a conversion factor.

recharge_annual_statistics—This file contains mean values of the same water-balance components reported in the recharge_daily_statistics file. The values represent the annual sum of the daily mean values. The fields for continuous frozen ground index (CFGI), snow cover, and accumulated potential water loss are of questionable value because they represent the sum of summed values. The soil-moisture field makes sense if it is divided by the number of days in a year.

The fourth file, the log file, is written each time the SWB code is run. This file not only contains a detailed transcript of model execution, but also records all SWB settings and informational messages. In the event that the SWB code fails to run as expected, the last few lines of the log file should provide clues regarding the reason for the failure.

Grid and Image Files

There are currently 26 internal and state variables that can be output either as Arc ASCII or Surfer grid files, image files, or both. Output may be generated for any of these variables as daily, monthly, or yearly values. The available output variables are described in table 11.

In addition, several grid files are output automatically when the SWB model runs. Under the output\future directory, two files per year are written with names similar to final_snow_cover_YYYY.asc and final_pct_sm_YYYY.asc, where YYYY is the calendar year of the simulation run. These files represent the snow cover (as snow water equivalent, in inches) and the soil moisture (in percentage of maximum water capacity) on December 31 of the simulated year. These files may be used with the INITIAL_SOIL_MOISTURE and INITIAL_SNOW_COVER options described below to supply more realistic model initial conditions.

The code also generates a final image file and grid file of the mean recharge over the entire simulation period (from year YYYY1 to YYYY2), located in the output and image subdirectories and labeled MEAN_RECHARGE_YYYY1-YYYY2.*. In a multiyear run, the averaging period for these files may be changed by using the STATS_START_YEAR and STATS_END_YEAR control-file directives. The default output is the mean recharge for all years in the simulation.

Table 11.   Description of available output variables.

Output variable name	Description
GROSS_PRECIP	Gross precipitation, echoing the input precipitation fed into the SWB model.
SNOWFALL	Reflects the amount of NET_PRECIP that falls when (TMAX-TMIN) 2/3 +TMIN exceeds the freezing point of water (32 degrees Fahrenheit).
SNOWCOVER	Tracks the cumulative difference between SNOWFALL and SNOWMELT on a daily basis. This cumulative difference can be reported as monthly annual value, but the result is meaningless. Note that snow cover is expressed as snow water equivalent, in inches.
CFGI	Continuous frozen ground index, in degrees–Celsius days.
CHG_IN_SNOW_COV	Change in snowcover, expressed as snow water equivalent, in inches.
SNOWMELT	Snowmelt, expressed as snow water equivalent, in inches.
MIN_TEMP	Minimum daily air temperature, echoing the input minimum air temperature supplied to SWB.
MAX_TEMP	Maximum daily air temperature, echoing the input maximum air temperature supplied to SWB.
AVG_TEMP	Mean daily air temperature, echoing the input mean air temperature supplied to SWB.
INTERCEPTION	Interception of precipitation, in inches.
NET_PRECIP	Gross precipitation minus interception, in inches.
INFLOW	Surface flow supplied to a grid cell from upslope, in inches.
OUTFLOW	Surface flow leaving a grid cell to an adjacent downslope cell, in inches.
RUNOFF_OUTSIDE	Surface flow that flows out of the model domain, or has reached a cell identified as open water, in inches.
REJECTED_RECHARGE	Recharge in excess of the daily maximum recharge rate, in inches.
NET_INFLOW	Sum of net precipitation, snowmelt, and inflow terms to a cell, in inches.
NET_INFIL	Net inflow, minus the runoff outside and outflow terms, in inches. This is the water that is actually available to enter soil storage.
POT_ET	Potential evapotranspiration, as calculated by the currently selected algorithm, in inches.
ACT_ET	Actual evapotranspiration, in inches. Potential evapotranspiration is limited by the amount of soil moisture available for evaporative loss given the soil accumulated potential water loss and the root zone depth of the cell.
P_MINUS_PET	This term is called "precipitation minus potential ET" in Thornthwaite and Mather (1957). Actual value is the net infiltration term minus the potential evapotranspiration term for a cell, in inches.
SM_DEFICIT	Daily soil moisture deficit, in inches. This is the amount by which evapotranspiration exceeds net infiltration of water.
SM_SURPLUS	Daily soil moisture surplus, in inches. This value represents the amount by which infiltrated water exceeds the maximum water capacity of the soil.
SM_APWL	Accumulated potential water loss, in inches. This is a running sum of the precipitation minus potential ET value for periods where ET exceeds precipitation.
SOIL_MOISTURE	Soil moisture, in inches.
CHG_IN_SOIL_MOIS	Change in soil moisture, in inches.
RECHARGE	Recharge, in inches.

Program Options and the Control File

This section describes the program options used in an SWB simulation. The options are contained in a text file. It is best to run the SWB code in a command window; in Windows XP, the command window may be started by pressing **Start**, selecting **Run**, and typing cmd.exe. The command to run SWB should be executed from the base of the directory structure as described elsewhere in this document. The name of a control file must be specified in order to begin execution of the SWB code.

Starting SWB

```
swb control_filename
```

Options are described below, and are presented in two groups: required and optional. Any text that begins with a pound sign (#) in the control file is ignored so that the user can include comments regarding data sources and program options within the control file. The actual required syntax of the options is shown in bold text.

Required Control-File Entries

MODEL DOMAIN DEFINITION [REQUIRED]

The first program option defines the extent of the model domain. Units of meters are assumed; an optional control-file statement (GRID_LENGTH_UNITS) may be used to specify that units of feet are used instead. If any subsequent input grid fails to match the specified model domain exactly, the program will end.

Example:

```
               Lower LH Corner    Upper RH Corner   Grid
              |_____|  |_____| Cell
    NX  NY  X0          Y0         X1         Y1      Size
GRID 344 320 528093.87 274821.57 562493.87 306821.57 100.0
```

In the example shown above, NX is the number of grid cells in the x direction, NY is the number of grid cells in the y direction, X0 and Y0 are the coordinates for the lower left-hand corner of the grid, and X1 and Y1 are the coordinates for the upper right-hand corner of the grid.

GROWING SEASON START AND END [REQUIRED]

This flag controls when the growing season is considered to begin and end (expressed as the day of year) and whether or not the problem is in the Northern hemisphere (possible values: TRUE/FALSE).

Example:

```
GROWING_SEASON 133 268 TRUE
```

This option affects which interception value (growing season or dormant season) and which antecedent runoff condition thresholds (table 1) apply to a grid cell.

PRECIPITATION [REQUIRED]

This control statement either indicates that precipitation at a single station is to be used or, alternatively, specifies a file type and a file prefix that the SWB code will associate with daily gridded precipitation values. The required syntax for gridded precipitation data is PRECIPITATION FILE_FORMAT precip_prefix.

Example – single station:

PRECIPITATION SINGLE_STATION

- or -

Example – Arc ASCII daily grid:

PRECIPITATION ARC_GRID climate\precip\PRCP

- or -

Example – Surfer daily grid:

PRECIPITATION SURFER climate\precip\PRCP

The naming convention used when specifying the gridded data must adhere to the following template: prefix_YYYY_MM_DD.suffix, where in this example prefix takes the value PRCP and YYYY, MM, and DD are the four-digit year, two-digit month, and two-digit day associated with the gridded data. The suffix expected is the same as that specified for gridded output; if unspecified, the suffix is asc by default. Note that the file prefix may specify one or more relative subdirectory names separated by backslashes (\). Precipitation grids are expected to be of data type real, with units of inches per day.

TEMPERATURE [REQUIRED]

This control statement either indicates that temperature at a single station is to be used or, alternatively, specifies a file type and file prefixes associated with daily gridded maximum and minimum temperature values. The required syntax for gridded temperature data is: TEMPERATURE FILE_FORMAT TMAXprefix TMINprefix.

Example – single station:

TEMPERATURE SINGLE_STATION

- or -

Example – Arc ASCII daily grid:

TEMPERATURE ARC_GRID climate\temp\tmax climate\temp\tmin

- or -

Example – Surfer daily grid:

TEMPERATURE SURFER climate\temp\tmax climate\temp\tmin

The naming convention used when specifying the gridded data must adhere to the following template: prefix_YYYY_MM_DD.suffix, where the prefix is as specified in the above control statement and YYYY, MM, and DD are the four-digit year, two-digit month, and two-digit day associated with the gridded data. The suffix expected is the same as that specified for gridded output; if unspecified, the suffix is "asc" by default. Note that the file prefix may specify one or more relative subdirectory names separated by backslashes (\). Temperature grids are expected to be of data type real, and in units of degrees Fahrenheit.

FLOW DIRECTION [REQUIRED *ONLY* IF FLOW ROUTING IS ENABLED]

Arc ASCII or Surfer grid of D8 flow directions. The flow-direction grid must be of data type integer.

Example:

```
FLOW_DIRECTION ARC_GRID input\new_fl_dir.asc
```

SOIL GROUP [REQUIRED]

Arc ASCII or Surfer grid of hydrologic soil groups; values must correspond to the hydrologic soil groups contained in the land-use lookup table. The soil group must be of data type integer.

Example:

```
SOIL_GROUP ARC_GRID input\soil_hyd_grp.asc
```

LAND-USE CLASSIFICATION [REQUIRED]

Arc ASCII or Surfer grid of land-use/land-cover classification values; the values must correspond to the land-use/land-cover values contained in the land-use lookup table. The land-use grid must be of data type integer.

Example:

```
LAND_USE ARC_GRID input\land_use.asc
```

LAND-USE LOOKUP TABLE [REQUIRED]

The name of the land-use lookup table must be specified within the control file.

Example:

```
LAND_USE_LOOKUP_TABLE std_input\LU_lookup.txt
```

AVAILABLE SOIL—WATER CAPACITY [REQUIRED]

The model uses soil information, together with land-cover information, to calculate surface runoff and assign a maximum soil-moisture holding capacity to each grid cell. The soil-group grid may be used along with the values included in table 7 to produce this grid. Available water capacity is expected to be of data type real and is expressed in units of inches of water per foot of soil.

Example:

```
WATER_CAPACITY ARC_GRID input\avail_water_cap.asc
```

SOIL-MOISTURE—ACCOUNTING METHOD [REQUIRED]

The model currently only contains one soil-moisture-accounting calculation option: Thornthwaite-Mather (1948, 1957). The Thornthwaite-Mather soil-moisture-retention tables are included in the standard table soil-moisture-retention-extended.grd.

Example:

```
SM T-M std_input\soil-moisture-retention-extended.grd
```

INITIAL SOIL MOISTURE [REQUIRED]

Initial soil moisture can be specified either as a single constant value or as a grid of values. Initial soil moisture is expressed as a percentage saturation of the available water capacity (0-100 percent). If supplied as gridded data, the initial soil-moisture grid is expected to be of data type real.

Example:

```
INITIAL_SOIL_MOISTURE CONSTANT 100
```

- or -

```
INITIAL_SOIL_MOISTURE ARC_GRID input\initial_pct_sm_dec_31_1999.asc
```

INITIAL SNOW COVER [REQUIRED]

Initial snow cover can be specified either as a single constant value or as a grid of values. Initial snow cover is expressed as a water-equivalent value (inches of water). If supplied as gridded data, the initial-snow-cover grid is expected to be of data type real.

Example:

```
INITIAL_SNOW_COVER CONSTANT 0
```

- or -

```
INITIAL_SNOW_COVER ARC_GRID input\initial_snow_cover_dec_31_1999.asc
```

RUNOFF CALCULATION AND ROUTING METHOD [REQUIRED]

Currently, only one runoff calculation method is available: the NRCS curve number method (Cronshey and others, 1986). This calculation method must be specified by including the letters C-N after the keyword RUNOFF.

Three methods are available for the routing of surface water through the model domain.

The ITERATIVE method is based on the original VBA code solution method, in which water is iteratively moved across the entire grid until all water has either infiltrated or left the grid via surface flow. The DOWNHILL method involves a pre-simulation step in which the model grid cells are sorted upslope to downslope. Runoff is calculated in a single iteration for each time step over the entire model domain, proceeding from the upstream cells to the downstream cells. There should be no difference between the two routing methods except that the DOWN-HILL method executes much more quickly.

Specifying the method as NONE disables routing altogether. Any cell outflow is assumed to find its way to a surface-water feature and exit the model domain. Outflow under this option is tracked as RUNOFF_OUTSIDE.

The SWB code writes a binary file to disk after the first pass through the DOWNHILL iteration method. This binary file is read in for subsequent SWB model runs, eliminating the potentially time-consuming task of grid-cell sorting. Any changes to the grid extent will trigger an error message reminding the user to delete the file swb_routing.bin before rerunning the SWB code.

Example:

```
RUNOFF C-N ITERATIVE
```

- or -

```
RUNOFF C-N DOWNHILL
```

- or -

```
RUNOFF C-N NONE
```

EVAPOTRANSPIRATION METHOD [REQUIRED]

The model implements several methods for estimating potential evapotranspiration, specifically, the Thornthwaite-Mather, Jensen-Haise, Blaney-Criddle, Turc, and Hargreaves-Samani methods. Note that, given the same root-zone depths, the Thornthwaite-Mather method will produce lower estimates of potential evapotranspiration (and thus, higher estimates of recharge) than the other methods (Vörösmarty and others, 1998). The Hargreaves-Samani method is the only one suitable for use with gridded precipitation and air-temperature data.

The complete list of possible program options for specifying an ET calculation method is given below.

Example: Thornthwaite-Mather

```
ET T-M latitude
```

- or -

Example: Jensen-Haise

```
ET J-H latitude albedo a_s b_s
```

- or -

Example: Blaney-Criddle

```
ET B-C latitude
```

- or -

Example: Turc

```
ET TURC latitude albedo a_s b_s
```

- or -

Example: Hargreaves

```
ET HARGREAVES southerly_latitude northerly_latitude
```

Values must be entered for all specified options. In the absence of more specific information, a reasonable value of the albedo for the Jensen-Haise and Turc methods is 0.23; similarly, a_s may be set to 0.25 and b_s set to 0.5. The coefficients a_s and b_s are used in the Angstrom formula for estimation of daily solar radiation (Allen and others, 1998). The term a_s expresses the fraction of extraterrestrial radiation that reaches earth on overcast days; b_s expresses the additional fraction of extraterrestrial radiation that reaches earth on clear days.

SOLVE [REQUIRED]

This option begins the actual recharge calculation. The name of the single-station climate time-series data file is required.

Example:

```
SOLVE bec1999v2.txt
```

Note that a simulation for more than 1 year is made possible by simply including additional SOLVE statements.

Example:

```
SOLVE MSN_1999.txt

SOLVE MSN_2000.txt

SOLVE MSN_2001.txt

... etc ...
```

If gridded temperature and precipitation data are supplied to the SWB model and the Hargreaves-Samani (1985) ET calculation method is selected for use, the single-station climate time-series file may be eliminated altogether. In this case, the starting and ending year of the simulation must be supplied.

Example:

```
SOLVE_NO_TS_FILE 1983 1992
```

END-OF-JOB FLAG [REQUIRED]

This program option triggers actions to write grids to disk, calculate statistics, and de-allocate memory.

Example:

```
EOJ
```

Optional Control-File Entries

The control-file entries discussed above are the only ones required to run the SWB model. However, there are additional control-file entries available that change the method of calculation or enable additional output options. This section describes optional control-file entries that may be used.

ANSI COLORED TEXT [OPTIONAL]

If you have access to a terminal program such as rxvt, the SWB model can generate screen output with color coding for positive and negative values (possible values: TRUE/FALSE). The rxvt package can be installed on a Windows PC as an option along with the Cygwin Unix emulation package (www.cygwin.com).

Example:

```
ANSI_COLORS TRUE
```

OUTPUT SUPPRESSION [OPTIONAL]

In order to speed model runtimes, certain types of text messages normally printed to the screen and/or disk may be suppressed. SUPPRESS_SCREEN_OUTPUT will turn off the detailed mass-balance information that is normally printed to the screen for each daily timestep. SUPPRESS_DAILY_FILES will prevent detailed mass balance from being written to disk as recharge_daily_statistics.csv, recharge_annual_report.csv, and recharge_daily_report.csv. SUPPRESS_DISLIN_MESSAGES will prevent the progress messages normally generated by the DISLIN graphics library from being written to the screen.

Example:

```
SUPPRESS_SCREEN_OUTPUT
#SUPPRESS_DAILY_FILES
SUPPRESS_DISLIN_MESSAGES
```

ADJUSTED WATER CAPACITY [OPTIONAL]

The model will calculate the maximum available water capacity from the base soil-water-capacity grid and the land-use grid, using the rooting-depth values as specified in the land-use lookup table. Alternatively, the adjusted water capacity may be calculated independently of the model and read in as a real-number ASCII grid. If this is done, internal calculation of the rooting depth and resulting adjusted water capacity is disabled in the model.

Example:

```
ADJUSTED_WATER_CAPACITY ARC_GRID input\MAX_SM_STORAGE.asc
```

CONTINUOUS FROZEN GROUND THRESHOLD VALUES [OPTIONAL]

The upper and lower continuous-frozen-ground indices may be set with the UPPER_LIMIT_CFGI and LOWER_LIMIT_CFGI statements. As discussed elsewhere, these values define the boundaries between completely frozen soil (the upper limit) and completely unfrozen soil (the lower limit). The CFGI threshold values are expressed in degree-Celsius-days.

Example:

```
UPPER_LIMIT_CFGI 83
LOWER_LIMIT_CFGI 56
```

INITIAL FROZEN GROUND INDEX [OPTIONAL]

This statement sets the initial (year 1) continuous-frozen-ground index. This may be supplied as a constant (CONSTANT) or as a gridded data set (ARC_GRID or SURFER).

Example:

```
INITIAL_FROZEN_GROUND_INDEX CONSTANT 100.0
```

Example:

```
INITIAL_FROZEN_GROUND_INDEX ARC_GRID input\INIT_CFGI.asc
```

INITIAL ABSTRACTION METHOD [OPTIONAL]

The method for calculating the initial abstraction within the NRCS curve number method may be specified in one of two ways:

1. TR-55: I_a is assumed equal to 0.2 S,

2. Woodward and others (2003): I_a is assumed equal to 0.05 S.

If the Hawkins method is used, curve numbers are adjusted as given in equation 9 of Woodward and others (2003). The overall effect should be to increase runoff for smaller precipitation events. This method has been suggested to be more appropriate to long-term simulation model applications.

Example:

```
#INITIAL_ABSTRACTION_METHOD TR55

INITIAL_ABSTRACTION_METHOD HAWKINS
```

OUTPUT GRID FILE FORMAT [OPTIONAL]

This option allows the user to choose the format of grid output. Currently two formats are supported: ESRI ASCII Grid, and Golden Software Surfer. The default is ARC_GRID.

Example:

```
OUTPUT_FORMAT ARC_GRID
```

- or -

```
OUTPUT_FORMAT SURFER
```

OUTPUT GRID FILENAME PREFIX [OPTIONAL]

This option sets the output grid filename prefix. If no value is supplied, output file names will begin with swb.

Example:

```
OUTPUT_GRID_PREFIX BlkErth
```

OUTPUT GRID FILENAME SUFFIX [OPTIONAL]

This option sets the output grid filename suffix. The default value is asc.

Example:

```
OUTPUT_GRID_SUFFIX txt
```

OUTPUT VARIABLES [OPTIONAL]

The code will write out gridded data files or image files (portable network graphics, Adobe Portable Document Format, or bitmap) for any of 24 internal and state variables simulated in the model (see table 11). The required syntax of this option is

```
OUTPUT_OPTIONS variable_name daily_output monthly_output annual_output
```

The list of valid output types is NONE, GRAPH, GRID, or BOTH.

For example, if one wished to produce output for actual evapotranspiration, the following would yield no daily output, graphical monthly output, and both gridded and graphical annual output.

Example:

```
OUTPUT_OPTIONS ACT_ET NONE GRAPH BOTH
```

Note that the monthly and annual output types represent the summation of the simulated daily values and may not have any physical meaning. For example, although annual gridded SNOWFALL output will represent the annual snowfall amount as water equivalent precipitation, the annual gridded SNOWCOVER output represents the summation of the daily amount of snow storage. It is unclear what utility this summation would have.

Also note that the graphical output is not publication quality but is rather included as a quick way to visualize changes in key variables over space and through time.

GRAPHICS PARAMETERS (DISLIN PARAMETERS) [OPTIONAL]

Plots of any of the variables listed in the "Output Variables" section above are created by use of the DISLIN plotting library (Michels, 2007). The plots created by the SWB code are quite basic. The plotting functionality in the SWB code is included primarily as a quick diagnostic tool for users.

SWB enables use of some of the more important DISLIN parameters. These parameters control how the DISLIN plotting library formats each plot. The syntax is DISLIN_PARAMETERS *SWB Output Variable Name*, followed on the next lines by one or more of the following statements: SET_Z_AXIS_RANGE, SET_DEVICE, SET_FONT, or Z_AXIS_TITLE.

The SET_Z_AXIS_RANGE statement allows the user to specify the range of values that will be plotted. Three sets of ranges may be specified for plots of daily, monthly, and annual values. The minimum, maximum, and increment size must be specified each time the SET_Z_AXIS_RANGE statement is used.

Example:

```
DISLIN_PARAMETERS RECHARGE

SET_Z_AXIS_RANGE DAILY 0 1.5 0.1

SET_Z_AXIS_RANGE MONTHLY 0 7 1.0

SET_Z_AXIS_RANGE ANNUAL 0 20 2.

SET_DEVICE PDF

SET_FONT Times-Bold

Z_AXIS_TITLE RECHARGE, IN IN.
```

In the second line of the example above, the plotting range for the RECHARGE output variable on a daily time scale is specified with a minimum of 0 in., a maximum of 1.5 in., and a plotting increment of 0.1 in.

The SET_DEVICE statement can be used to change the output file format. The default value is PNG, for Portable Network Graphics. This format is compact and is supported by most modern applications. Other possible output types include Windows Metafile (WMF), Adobe Postscript (PS), Adobe Encapsulated Postscript (EPS), Adobe Portable Document Format (PDF), Scalable Vector Graphics (SVG), and Windows Bitmap (BMP).

The SET_FONT statement can be used to alter the font used to annotate the plots. For plot device type PS, EPS, PDF, and SVG, the following subset of fonts should work:

Times-Roman	Courier
Helvetica	AvantGarde-Book
Helvetica-Narrow	Bookman-Light
NewCenturySchlbk-Roman	Palatino-Roman
NewCenturySchlbk-Italic	Palatino-Italic

For more details about supported fonts, see the DISLIN documentation (*http://www.dislin.de/*).

The statement Z_AXIS_TITLE sets the text that is used for the plot legend. Text does not need to be enclosed by quotes and may include punctuation.

ITERATIVE METHOD TOLERANCE [OPTIONAL]

The iterative method sometimes fails to converge for small solution tolerances (that is, less than 1.0E-6 change in calculated runoff in a cell from one iteration to the next). Increasing this value will improve convergence, but at the potential cost of also increasing mass-balance errors in the overall water balance. The default value is 1E-6.

Example:

```
ITERATIVE_METHOD_TOLERANCE 1.0E-4
```

UNITS OF LENGTH [OPTIONAL]

The SWB code is written with default units of meters to define the gridded model domain. If units of feet are desired instead, the GRID_LENGTH_UNITS statement may be used. The only place in the code where this is important is with regard to the mass balance as expressed in units of acre-feet. If the grid length units are feet while the code treats them as meters, there will be a corresponding error in the values of the mass-balance terms.

Example:

```
GRID_LENGTH_UNITS FEET
```

Test Problems

Test Case 1—Black Earth Creek, Dane and Iowa Counties, Wisconsin

This section compares the results of the current SWB code to those calculated by the Visual Basic for Applications (VBA) version of the code for a test case centered on Black Earth Creek, in Dane and Iowa Counties, Wis. The model domain for this test case coincides with a MODFLOW (Harbaugh and others, 2000) model domain with 100-m grid cells in an array of 356 columns and 331 rows, for a total of 117,836 cells.

Input Tables and Grids

Input data grids were created with available geographic information system (GIS) data. All grids were resampled such that their origin, extent, and grid-cell sizes conformed to the existing MODFLOW grid for the project area.

The land-use grid (fig. 7) for the test case was created by resampling and reclassifying the WISCLAND land-cover dataset for the project area (Wisconsin Department of Natural Resources, 1998). The resulting land-use grid was reclassified so that the land-use codes are consistent with the modified Anderson Level II scheme originally used by Dripps (2003).

The soils grid (fig. 8) is based on work originally done by the University of Wisconsin Land Information and Computer Graphics Facility (1988), obtained through the Dane County Land Information Office. This dataset contained the NRCS hydrologic soil group as a data element. The Dane County Digital Soil Survey has now largely been superseded by the Soil Survey Geographic Database products (SSURGO) available from the NRCS (*http://www.ncgc.nrcs.usda.gov/products/datasets/ssurgo/*).

The flow-direction grid (fig. 9) was generated from a 30-m digital elevation model of Wisconsin (U.S. Geological Survey, 2000a). Flow directions were determined using the ArcInfo GRID command **FLOWDIRECTION** to process the elevation data using the D8 algorithm (O'Callaghan and Mark, 1984).

The available-water-capacity grid (fig. 10) was created by assigning available-water-capacity values to the four soil types in the hydrologic soil group grid. Table 7 may be used to assist in the preparation of an available-water-capacity grid. Also, the newer SSURGO soil survey data may include more refined estimates of the available water capacity for each soil series.

Climatological data were obtained from the National Oceanic and Atmospheric Administration Co-op program. The closest high-quality data site relative to the Black Earth Creek study area is the station at the Dane County Regional Airport (National Climate Data Center Coop ID: 474961). All required and optional data elements were available for this station.

Simulation Details

The primary purpose of the Black Earth Creek test case was to compare the results of the SWB code to those produced by the original VBA code. Both models were run for the two-year period 1999–2000. The first year of simulation was used to initialize the model. Presumably, the initial conditions for soil moisture and snowfall at the start of 2000 are well approximated by the ending results for December 31, 1999.

The routing option RUNOFF was set to C-N ITERATIVE. The evapotranspiration option was set to T-M, which selects the Thornthwaite-Mather evapotranspiration routine.

Simulation Results

Comparisons between the two codes are shown in figures 11–14; results shown represent totals for the calendar year 1999. Basic summary statistics describing the differences were calculated by use of the R statistical package (R Core Development Team, 2008); these statistics are further broken down by land-use code. Difference statistics were calculated by subtracting the Visual Basic code result from the Fortran 95 code result at each grid cell. These differences are discussed below.

Net infiltration (fig. 11)—Differences between the two codes are small (mean difference = 0.00; median of differences = 0.00). The net infiltration values shown represent that portion of the total precipitation that is able to enter soil storage in a given year. The mean net infiltration for 1999 is estimated to be about 28 in/yr, or about 88 percent of precipitation that fell in this area during 1999.

Actual evapotranspiration (fig. 12)—Differences between the two codes are small (mean difference = 0.17; median of differences = 0.10). The Fortran 95 version values are slightly higher (median value = 20.59) than those calculated with the Visual Basic code (median value = 20.48). Under certain circumstances the Visual Basic code was observed to calculate a negative value for the potential evapotranspiration; negative values for potential evapotranspiration could ripple through the calculations and lower the actual evapotranspiration amounts. The mean actual evapotranspiration for 1999 is estimated at about 20.5 in/yr.

Ending soil moisture (fig. 13)—Differences between the two codes are small (mean difference = 0.08; median of differences = -0.03). Both codes were started with soil moisture values set to 100 percent of the maximum adjusted soil water capacity.

Recharge (fig. 14)—Differences between the two codes also are small (mean difference = -0.07; median of differences = -0.02). There are no obvious patterns in the differences with respect to the various land uses. The annual mean recharge estimated for 1999 is about 9.3 in. Analysis of streamflow records for the Black Earth Creek at Black Earth streamgage (USGS station number 05406500) for the years 1955–98 yields a base-flow estimate (assumed to be equivalent to recharge) of 9.5 in/yr (Gebert and others, 2007).

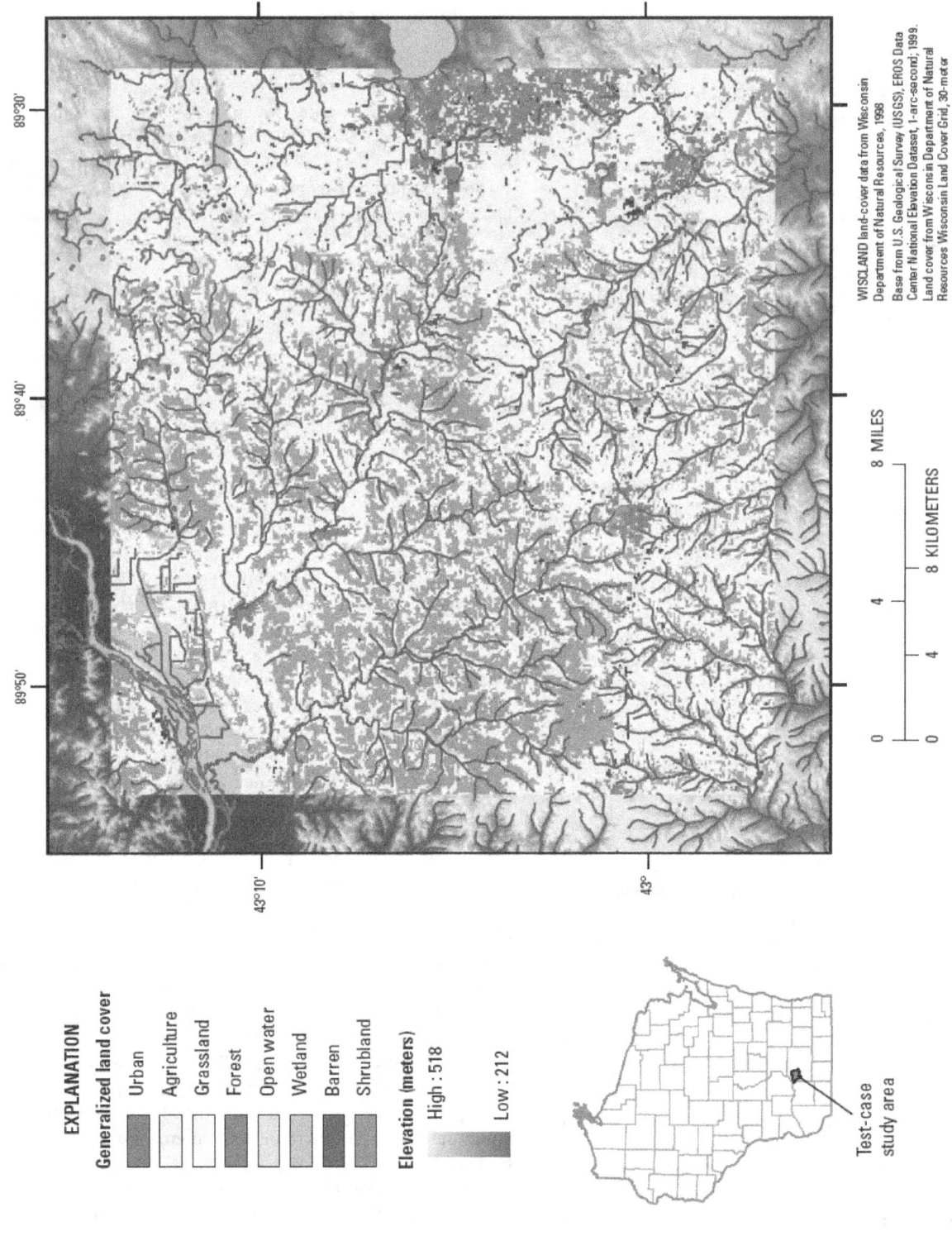

EXPLANATION

Generalized land cover

Urban
Agriculture
Grassland
Forest
Open water
Wetland
Barren
Shrubland

Elevation (meters)

High : 518

Low : 212

Test-case
study area

WISCLAND land-cover data from Wisconsin
Department of Natural Resources, 1998

Base from U.S. Geological Survey (USGS), EROS Data
Center National Elevation Dataset, 1-arc-second; 1999.
Land cover from Wisconsin Department of Natural
Resources Wisconsin Land Cover Grid, 30-meter
resolution, Madison, Wisconsin; 1999.

Figure 7.   Land use/land-cover classification for the Black Earth Creek test case.

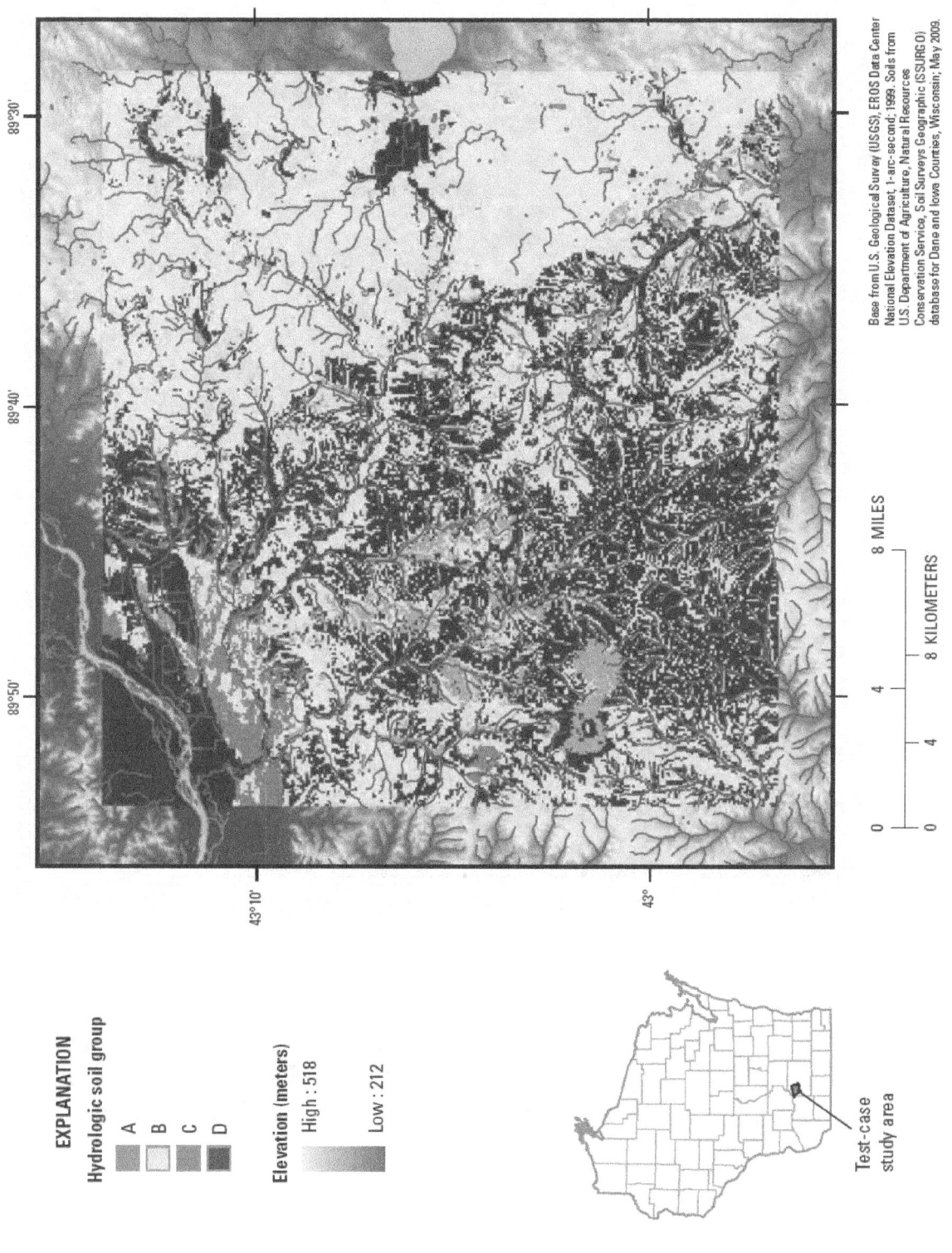

Base from U.S. Geological Survey (USGS), EROS Data Center
National Elevation Dataset; 1-arc-second; 1999. Soils from
U.S. Department of Agriculture, Natural Resources
Conservation Service, Soil Surveys Geographic (SSURGO)
database for Dane and Iowa Counties, Wisconsin; May 2009.

Figure 8.   Hydrologic soil groups for the Black Earth Creek test case.

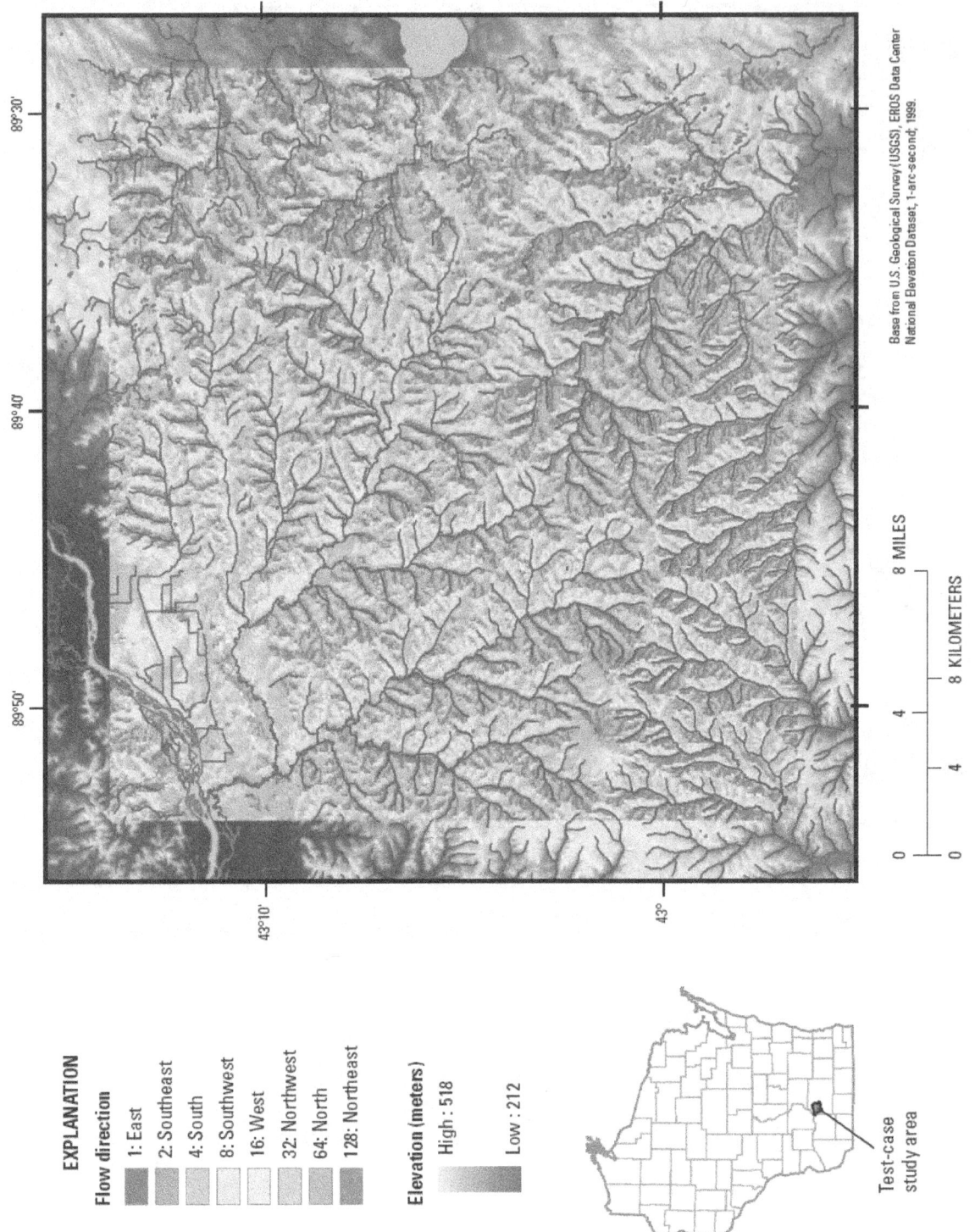

Figure 9. D8 flow-direction grid for the Black Earth Creek test case.

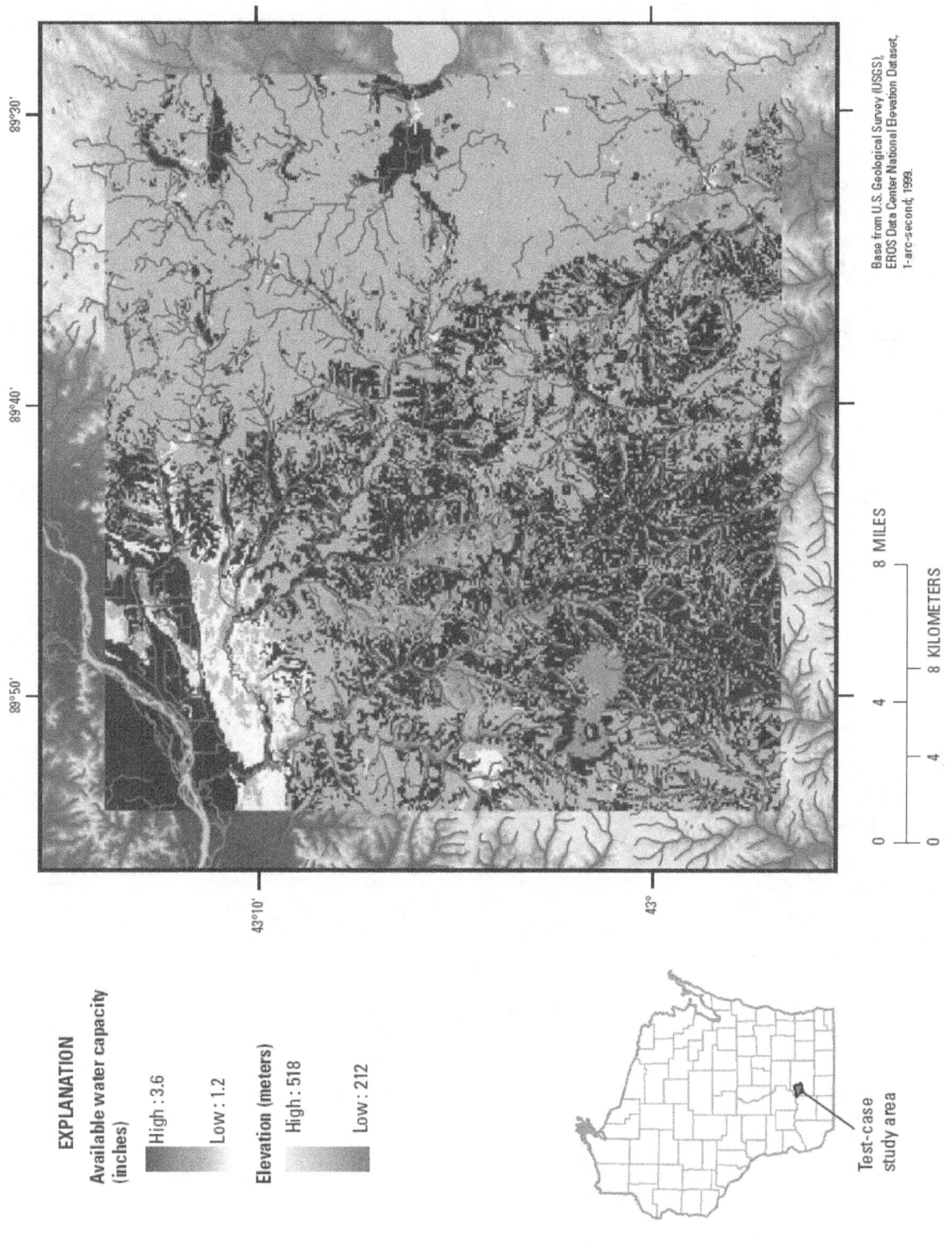

Base from U.S. Geological Survey (USGS),
EROS Data Center National Elevation Dataset,
1-arc-second, 1999.

Figure 10.   Available water capacity (AWC) for soils in the Black Earth Creek test case.

Net Infiltration – Black Earth Creek – 1999 (FORTRAN 95 code)

NET INFIL (in)   min: 4.91   max: 3466.77   mean: 28.01   median: 27.46

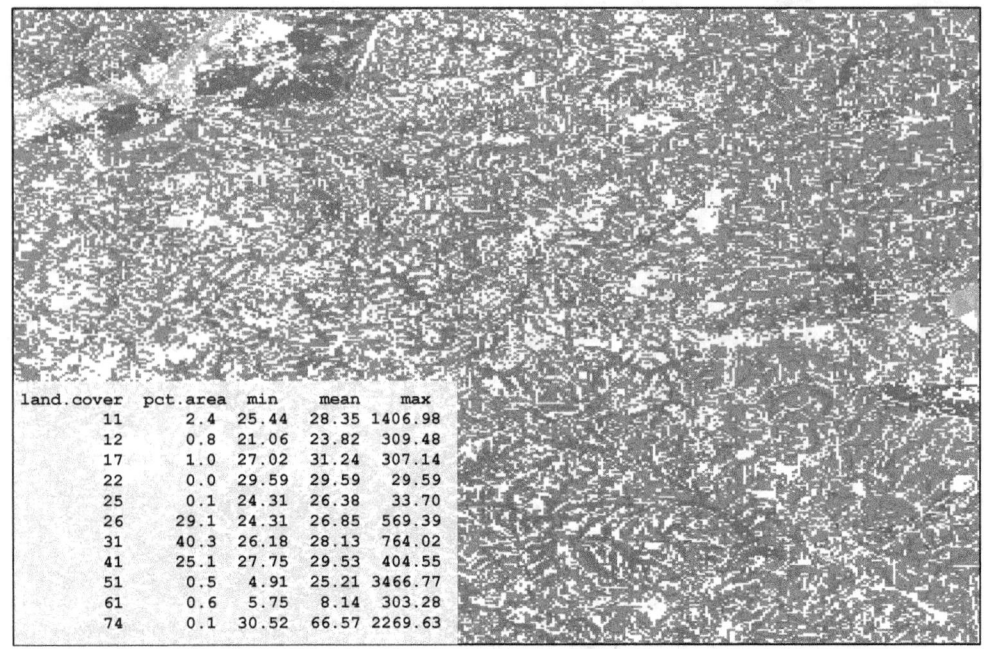

DIFFERENCE (F95 – VB): min: –71.75   max: 2.87   mean: –0.00   median: –0.00

Net Infiltration – Black Earth Creek – 1999 (VB code)

NET INFIL (in)   min: 3.95   max: 3538.53   mean: 28.01   median: 27.48

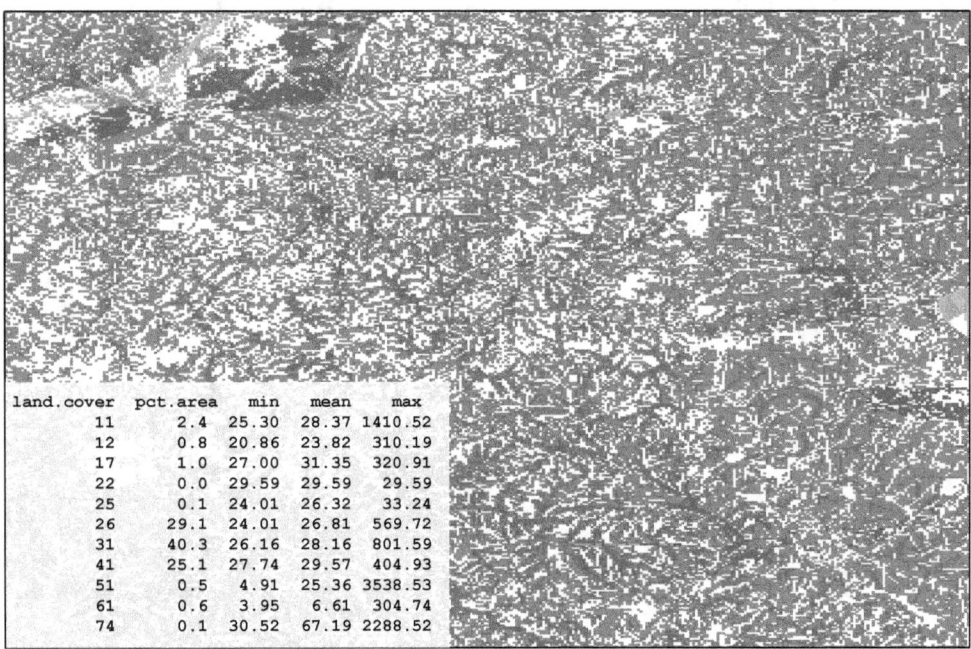

DIFFERENCE (F95 – VB): min: –71.75   max: 2.87   mean: –0.00   median: –0.00

EXPLANATION

Net infiltration,
in inches

26.67 to 28.00
25.33 to 26.67
24.00 to 25.33
22.67 to 24.00
21.33 to 22.67
20.00 to 21.33
18.67 to 20.00
17.33 to 18.67
16.00 to 17.33
14.67 to 16.00
13.33 to 14.67
12.00 to 13.33
10.67 to 12.00
9.33 to 10.67
8.00 to 9.33
6.67 to 8.00
5.33 to 6.67
4.00 to 5.33
2.67 to 4.00
1.33 to 2.67
0.00 to 1.33

Figure 11.   Comparison of Fortran 95 and Visual Basic versions of the SWB code—net infiltration, 1999.

Actual ET – Black Earth Creek – 1999 (FORTRAN 95 code)

ACT ET (in)   min: 2.05 max: 25.26 mean: 20.60 median: 20.59

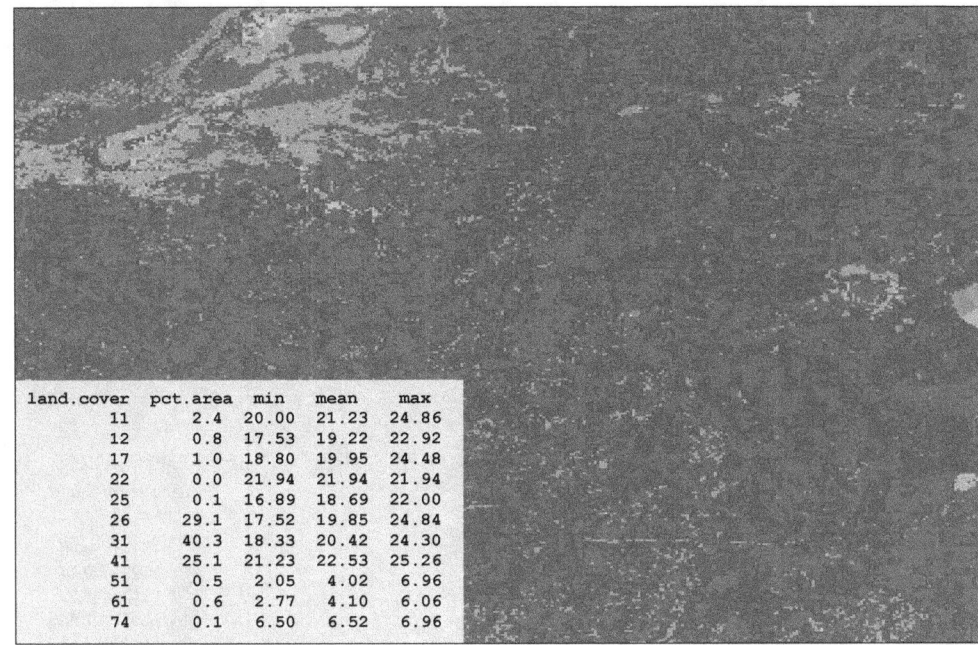

```
land.cover   pct.area   min    mean    max
        11        2.4  20.00  21.23  24.86
        12        0.8  17.53  19.22  22.92
        17        1.0  18.80  19.95  24.48
        22        0.0  21.94  21.94  21.94
        25        0.1  16.89  18.69  22.00
        26       29.1  17.52  19.85  24.84
        31       40.3  18.33  20.42  24.30
        41       25.1  21.23  22.53  25.26
        51        0.5   2.05   4.02   6.96
        61        0.6   2.77   4.10   6.06
        74        0.1   6.50   6.52   6.96
```

DIFFERENCE (F95 – VB): min: –1.51 max: 10.52 mean: 0.17 median: 0.10

EXPLANATION

Actual evapotranspiration,
in inches

- 23.81 to 25.00
- 22.62 to 23.81
- 21.43 to 22.62
- 20.24 to 21.43
- 19.05 to 20.24
- 17.86 to 19.05
- 16.67 to 17.86
- 15.48 to 16.67
- 14.29 to 15.48
- 13.10 to 14.29
- 11.90 to 13.10
- 10.71 to 11.90
- 9.52 to 10.71
- 8.33 to 9.52
- 7.14 to 8.33
- 5.95 to 7.14
- 4.76 to 5.95
- 3.57 to 4.76
- 2.38 to 3.57
- 1.19 to 2.38
- 0.00 to 1.19

Actual ET – Black Earth Creek – 1999 (VB code)

ACT ET (in)   min: 1.25 max: 25.25 mean: 20.43 median: 20.48

```
land.cover  pct.area    min    mean    max
        11        2.4  19.85  20.96  24.60
        12        0.8  17.30  19.09  22.85
        17        1.0  18.72  19.87  24.45
        22        0.0  21.82  21.82  21.82
        25        0.1  16.68  18.60  22.11
        26       29.1  17.23  19.72  24.78
        31       40.3  18.15  20.35  24.40
        41       25.1  11.16  22.16  25.25
        51        0.5   2.06   4.01   6.95
        61        0.6   1.25   3.02   6.06
        74        0.1   6.49   6.51   6.95
```

DIFFERENCE (F95 – VB): min: –1.51 max: 10.52 mean: 0.17 median: 0.10

Figure 12.   Comparison of Fortran 95 and Visual Basic versions of the SWB code—actual evapotranspiration, 1999.

Ending Soil Moisture – Black Earth Creek – 1999 (FORTRAN 95 code)

SOIL MOIST (in)   min: 0.00   max: 16.10   mean: 7.85   median: 7.61

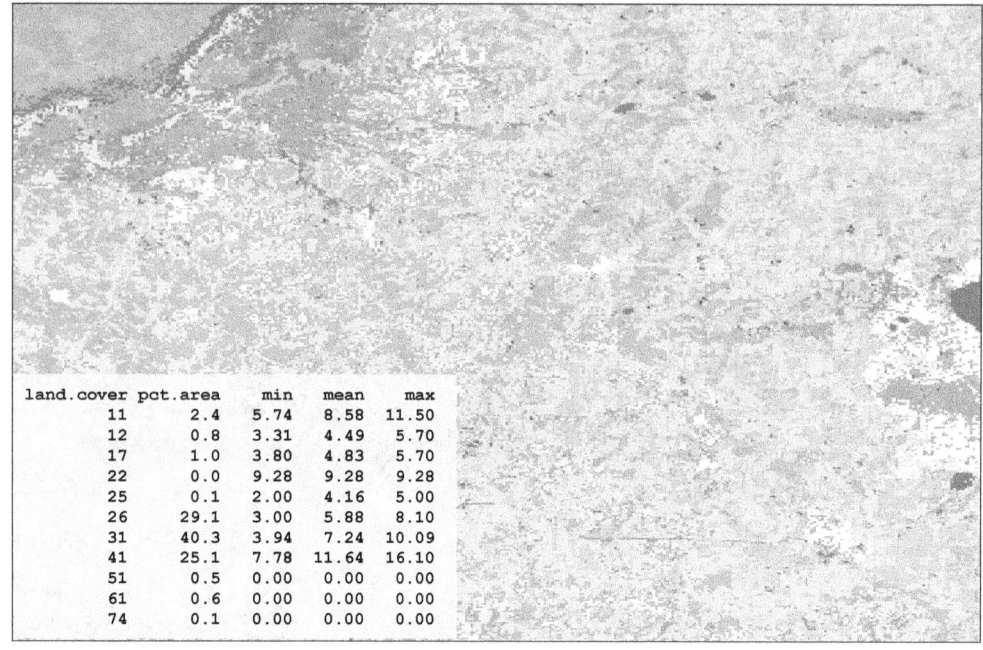

```
land.cover pct.area    min    mean     max
      11       2.4     5.74    8.58   11.50
      12       0.8     3.31    4.49    5.70
      17       1.0     3.80    4.83    5.70
      22       0.0     9.28    9.28    9.28
      25       0.1     2.00    4.16    5.00
      26      29.1     3.00    5.88    8.10
      31      40.3     3.94    7.24   10.09
      41      25.1     7.78   11.64   16.10
      51       0.5     0.00    0.00    0.00
      61       0.6     0.00    0.00    0.00
      74       0.1     0.00    0.00    0.00
```

DIFFERENCE (F95 – VB): min: –2.21   max: 0.90   mean: 0.08   median: –0.03

Ending Soil Moisture – Black Earth Creek – 1999 (VB code)

SOIL MOIST (in)   min: 0.00   max: 16.60   mean: 7.77   median: 7.64

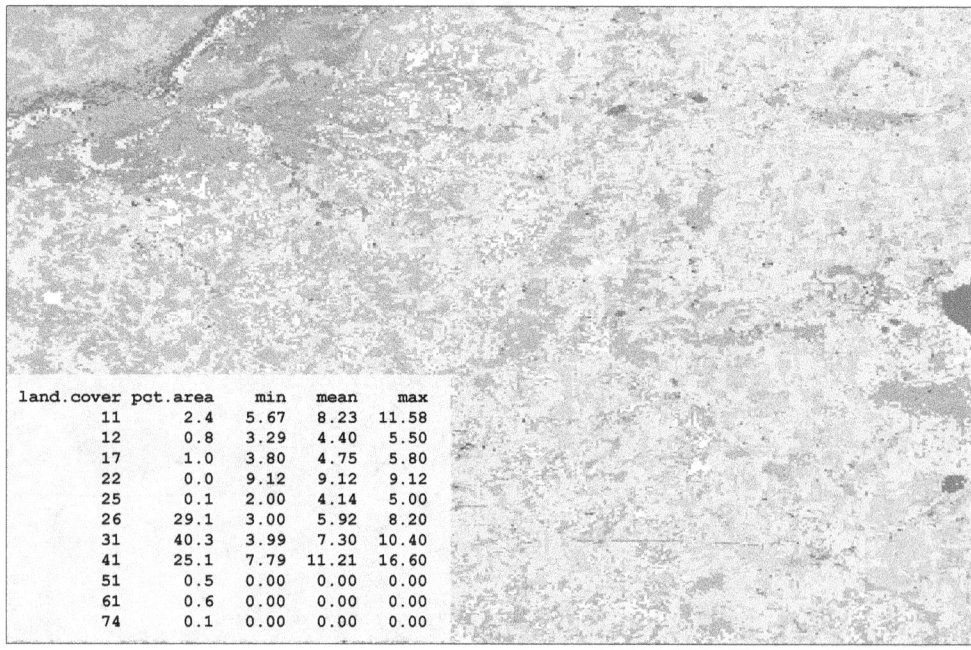

```
land.cover pct.area    min    mean     max
      11       2.4     5.67    8.23   11.58
      12       0.8     3.29    4.40    5.50
      17       1.0     3.80    4.75    5.80
      22       0.0     9.12    9.12    9.12
      25       0.1     2.00    4.14    5.00
      26      29.1     3.00    5.92    8.20
      31      40.3     3.99    7.30   10.40
      41      25.1     7.79   11.21   16.60
      51       0.5     0.00    0.00    0.00
      61       0.6     0.00    0.00    0.00
      74       0.1     0.00    0.00    0.00
```

DIFFERENCE (F95 – VB): min: –2.21   max: 0.90   mean: 0.08   median: –0.03

EXPLANATION

Ending soil moisture, in inches

	17.14 to 18.00
	16.29 to 17.14
	15.43 to 16.29
	14.57 to 15.43
	13.71 to 14.57
	12.86 to 13.71
	12.00 to 12.86
	11.14 to 12.00
	10.29 to 11.14
	9.43 to 10.29
	8.57 to 9.43
	7.71 to 8.57
	6.86 to 7.71
	6.00 to 6.86
	5.14 to 6.00
	4.29 to 5.14
	3.43 to 4.29
	2.57 to 3.43
	1.71 to 2.57
	0.86 to 1.71
	0.00 to 0.86

Figure 13.   Comparison of Fortran 95 and Visual Basic versions of the SWB code—ending soil moisture, 1999.

Recharge – Black Earth Creek – 1999 (FORTRAN 95 code)

RECHARGE (in)   min: 0.59   max: 3459.81   mean: 9.30   median: 8.83

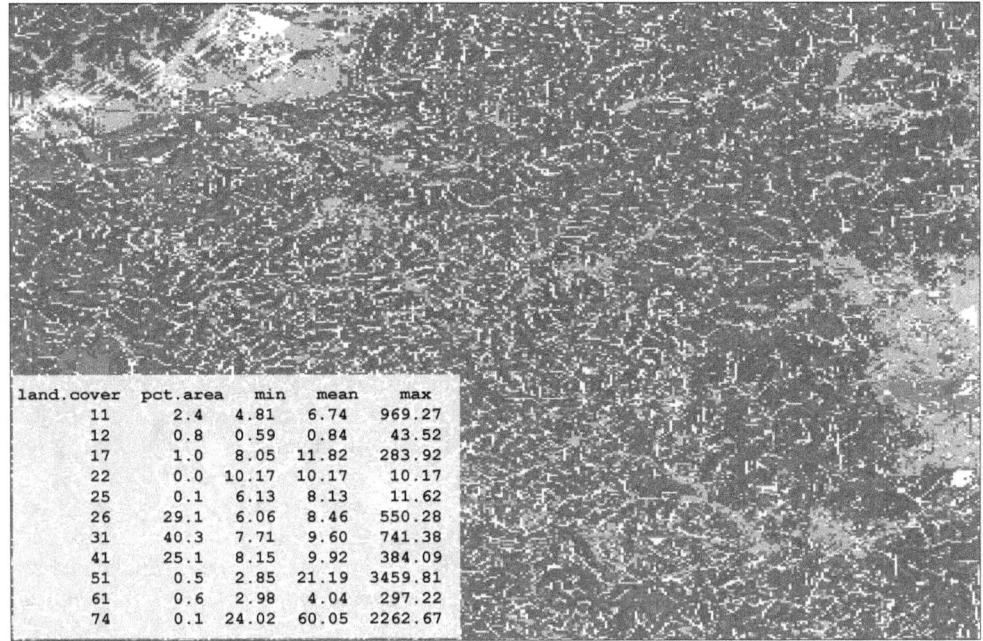

land.cover	pct.area	min	mean	max
11	2.4	4.81	6.74	969.27
12	0.8	0.59	0.84	43.52
17	1.0	8.05	11.82	283.92
22	0.0	10.17	10.17	10.17
25	0.1	6.13	8.13	11.62
26	29.1	6.06	8.46	550.28
31	40.3	7.71	9.60	741.38
41	25.1	8.15	9.92	384.09
51	0.5	2.85	21.19	3459.81
61	0.6	2.98	4.04	297.22
74	0.1	24.02	60.05	2262.67

DIFFERENCE (F95 – VB): min: –71.76   max: 2.49   mean: –0.07   median: –0.02

Recharge – Black Earth Creek – 1999 (VB code)

RECHARGE (in)   min: 0.59   max: 3531.57   mean: 9.37   median: 8.86

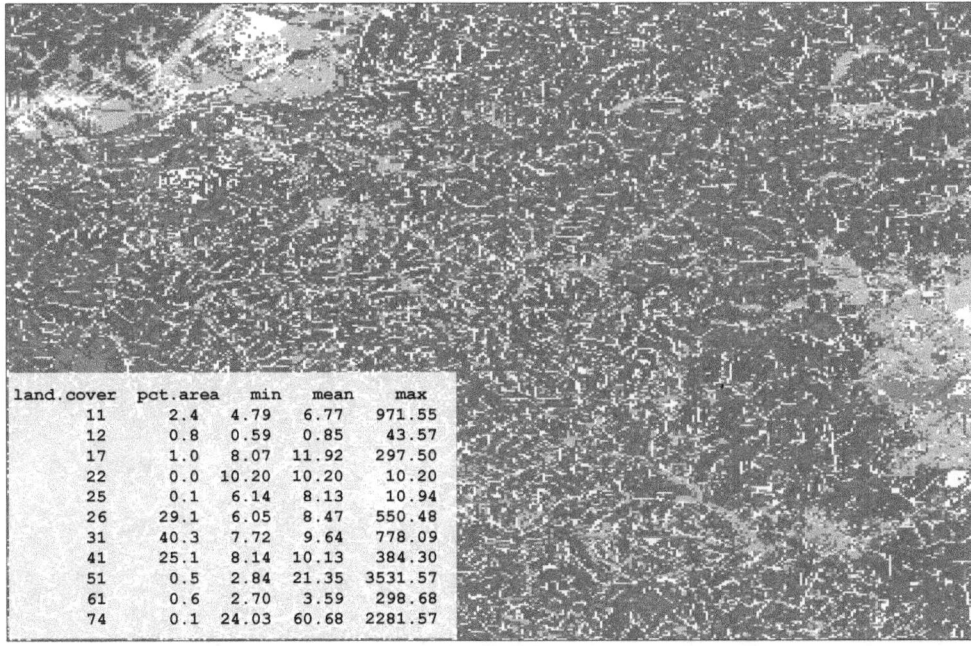

land.cover	pct.area	min	mean	max
11	2.4	4.79	6.77	971.55
12	0.8	0.59	0.85	43.57
17	1.0	8.07	11.92	297.50
22	0.0	10.20	10.20	10.20
25	0.1	6.14	8.13	10.94
26	29.1	6.05	8.47	550.48
31	40.3	7.72	9.64	778.09
41	25.1	8.14	10.13	384.30
51	0.5	2.84	21.35	3531.57
61	0.6	2.70	3.59	298.68
74	0.1	24.03	60.68	2281.57

DIFFERENCE (F95 – VB): min: –71.76   max: 2.49   mean: –0.07   median: –0.02

EXPLANATION

Recharge,
in inches

9.52 to 10.00
9.05 to 9.52
8.57 to 9.05
8.10 to 8.57
7.62 to 8.10
7.14 to 7.62
6.67 to 7.14
6.19 to 6.67
5.71 to 6.19
5.24 to 5.71
4.76 to 5.24
4.29 to 4.76
3.81 to 4.29
3.33 to 3.81
2.86 to 3.33
2.38 to 2.86
1.90 to 2.38
1.43 to 1.90
0.95 to 1.43
0.48 to 0.95
0.00 to 0.48

Figure 14.   Comparison of Fortran 95 and Visual Basic versions of the SWB code—recharge, 1999.

Test Case 2—Lake Michigan Basin

Application of the SWB code is demonstrated in this section for the Lake Michigan Drainage Basin, with a model grid of approximately 116,180 mi^2 covering parts of Michigan, Indiana, Illinois, Ohio, and Wisconsin, as well as a small part of Ontario. This test case illustrates the use of gridded precipitation and temperature data to capture the geographic variations in climate over large areas. The model domain consists of 129,560 grid cells in an array 316 cells wide and 410 cells high; each cell in the model domain has dimensions of 5,000 by 5,000 ft.

The input climate datasets used for this test case span the years 1989 to 2000, inclusive. Although not shown here, SWB was ultimately used to generate transient MODFLOW recharge arrays by use of gridded precipitation and air-temperature data spanning the period 1900 to 2000.

Input Tables and Grids

As in the Black Earth Creek test case, input data grids were created by use of available GIS data; grids were resampled consistent with the dimensions of the model domain (316 cells by 410 cells). The input grids at this scale were derived from national-level datasets, in contrast to the statewide and regional datasets used in the Black Earth Creek test case.

The land-use grid was derived from the 1992 National Land Cover Dataset (U.S. Geological Survey, 2000b). The resulting grid is shown in figure 15.

The classification used in this dataset is somewhat different from the modified Anderson Level II used in the original code (Dripps, 2003). Therefore, a new "standard" soil lookup table was developed to work with this land-use grid. The new lookup table contains a separate entry for each land-cover classification. Initial curve numbers were taken largely from the TR–55 publications (Cronshey and others, 1986) for the nearest matching land-cover classifications.

The soils grid was derived from mapping of glacial landforms by Fullerton and others (2003). The mapping units were generalized and grouped according to their approximate dominant grain size. Figure 16 shows the reclassified Fullerton hydrologic soil groups used in the test case.

The D8 flow-direction grid was derived from the 3-arc-second (90-m) "finished" Shuttle Radar Topography Mission (SRTM) data, obtained from *http://srtm.usgs.gov/*. The grid was resampled and the ArcInfo **FLOWDIRECTION** function was used to generate the D8 flow-direction grid (fig. 17). Ultimately, this grid was not used because the final SWB simulations were made with the flow routing turned off.

The available-water-capacity grid was created by applying table 7 to the textural classes associated with the hydrologic soil groups shown in figure 16. The resulting grid of available water capacity is shown in figure 18.

In all, 4,380 separate grid files were created for precipitation and maximum and minimum temperature data corresponding to each of the days of model simulation between 1989 and 2000, inclusive. More than 800 cooperative climate stations were included in the dataset used to generate the daily climate grids. The Fields library for the R statistical package was used to create a thin-plate spline of the daily precipitation values (Fields Development Team, 2006). The interpolated values were then written to disk as a series of Arc ASCII grid files. An example of the interpolated precipitation data is shown in figure 19; the data presented here represent the annual mean gross precipitation for the Lake Michigan Basin for the years 1990–2000, inclusive. Details regarding the construction of the climatological grid data files are discussed in Appendix 2.

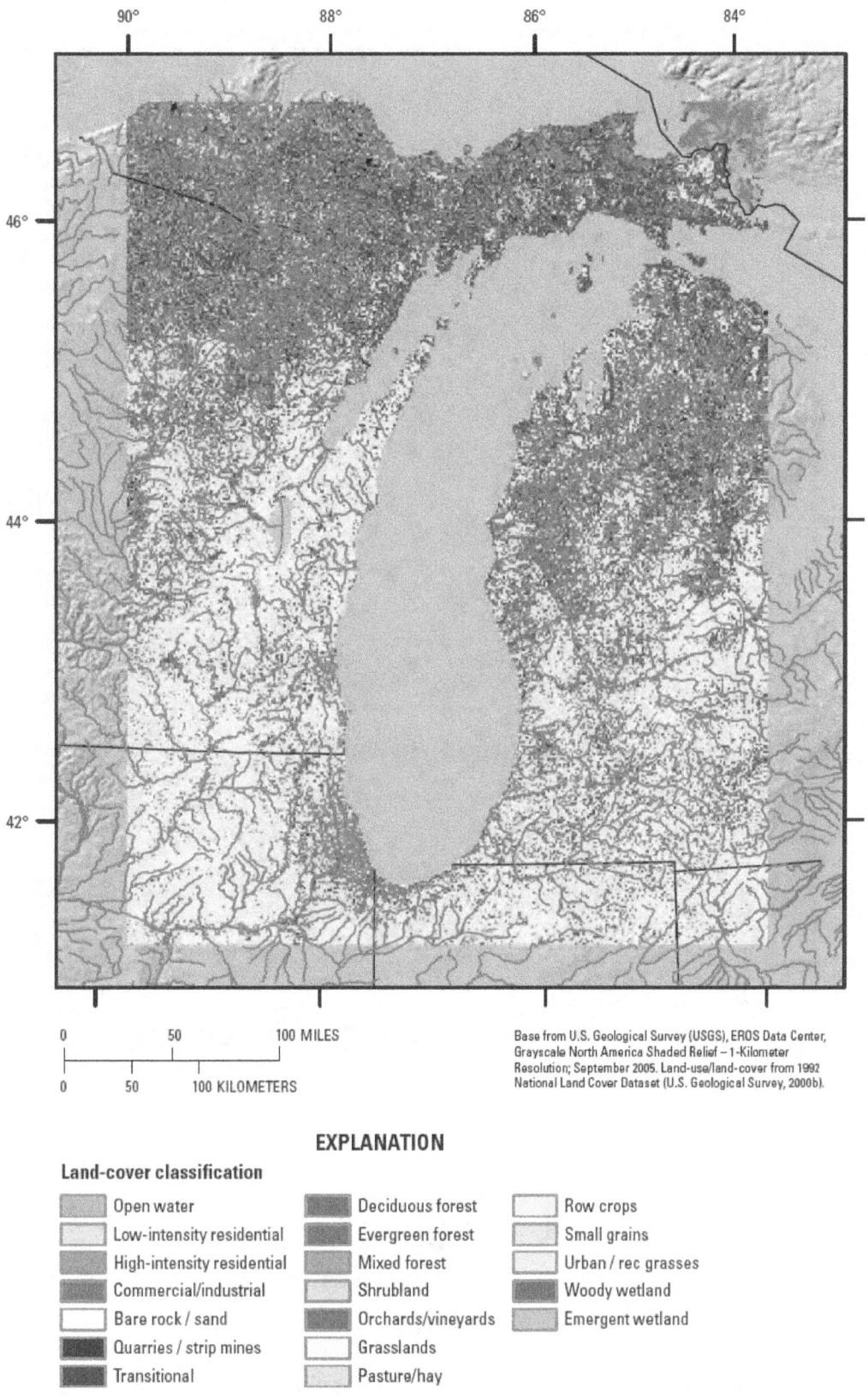

Base from U.S. Geological Survey (USGS), EROS Data Center,
Grayscale North America Shaded Relief – 1-Kilometer
Resolution; September 2005. Land-use/land-cover from 1992
National Land Cover Dataset (U.S. Geological Survey, 2000b).

EXPLANATION

Land-cover classification

Open water

Low-intensity residential

High-intensity residential

Commercial/industrial

Bare rock / sand

Quarries / strip mines

Transitional

Deciduous forest

Evergreen forest

Mixed forest

Shrubland

Orchards/vineyards

Grasslands

Pasture/hay

Row crops

Small grains

Urban / rec grasses

Woody wetland

Emergent wetland

Figure 15.   Land-cover classification for the Lake Michigan Basin test case.

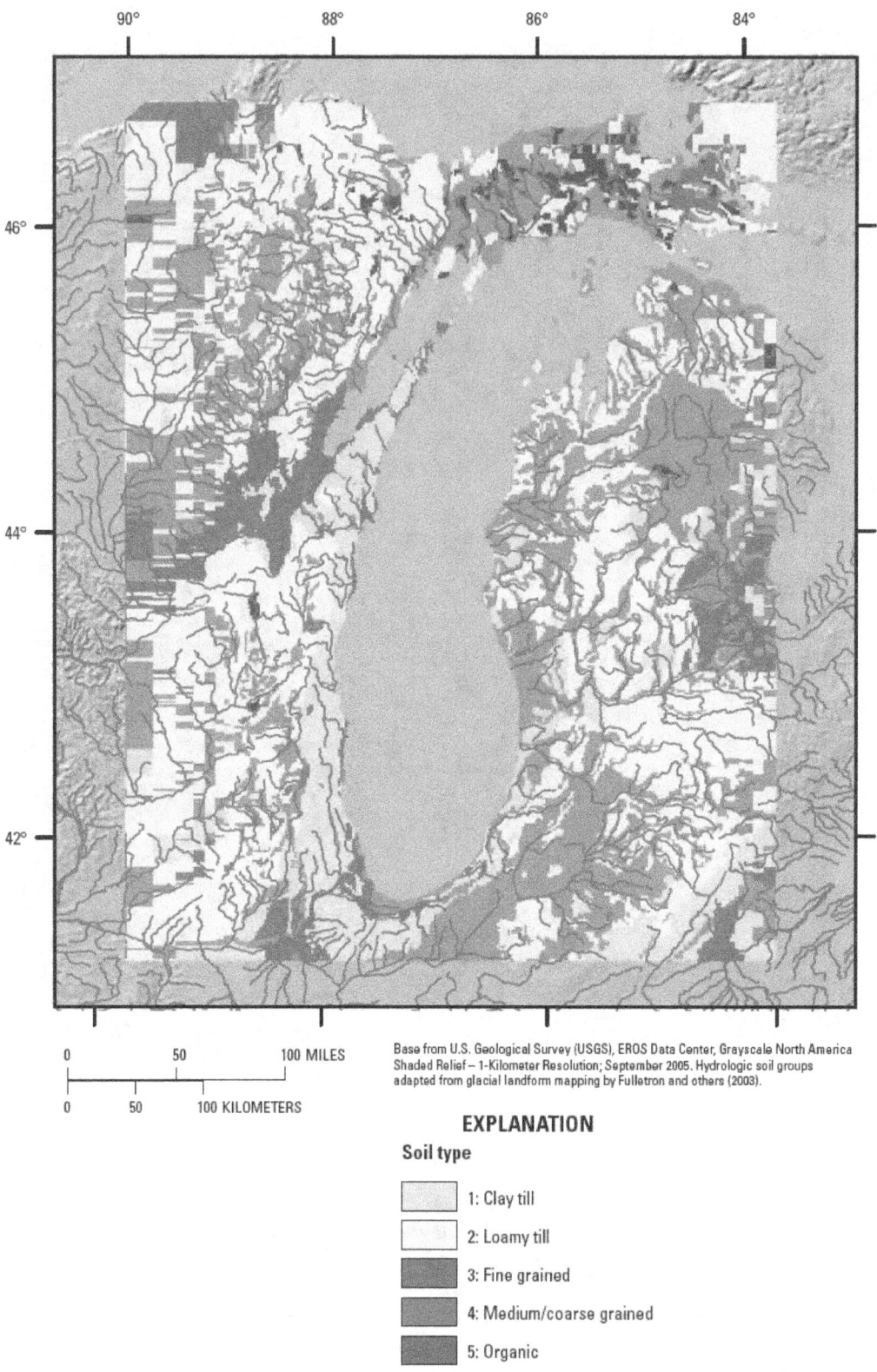

Base from U.S. Geological Survey (USGS), EROS Data Center, Grayscale North America
Shaded Relief – 1-Kilometer Resolution; September 2005. Hydrologic soil groups
adapted from glacial landform mapping by Fullerton and others (2003).

EXPLANATION

Soil type

1: Clay till

2: Loamy till

3: Fine grained

4: Medium/coarse grained

5: Organic

Figure 16.   Hydrologic soil groups for the Lake Michigan Basin test case.

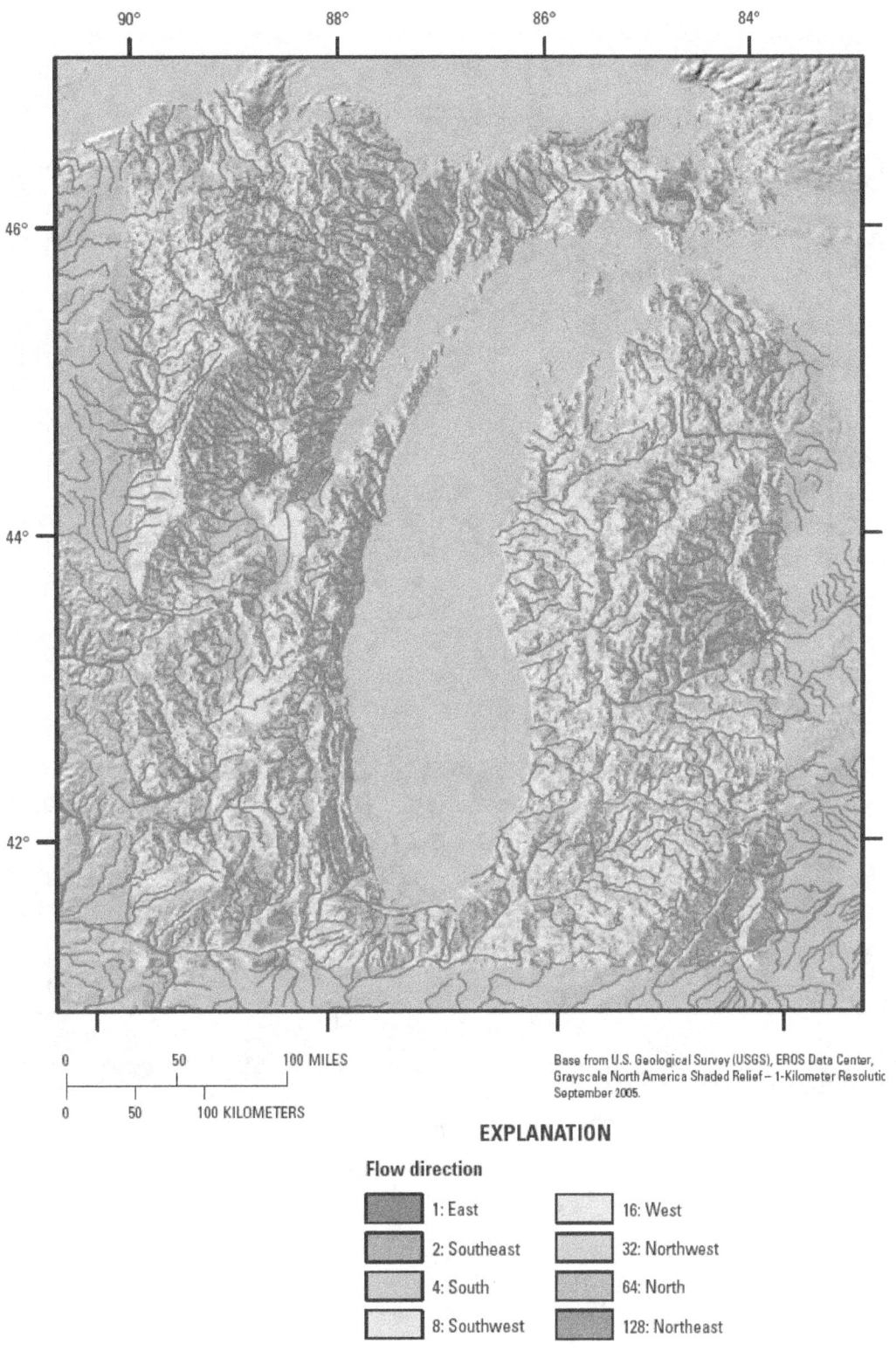

Figure 17. D8 flow-direction grid for the Lake Michigan Basin test case.

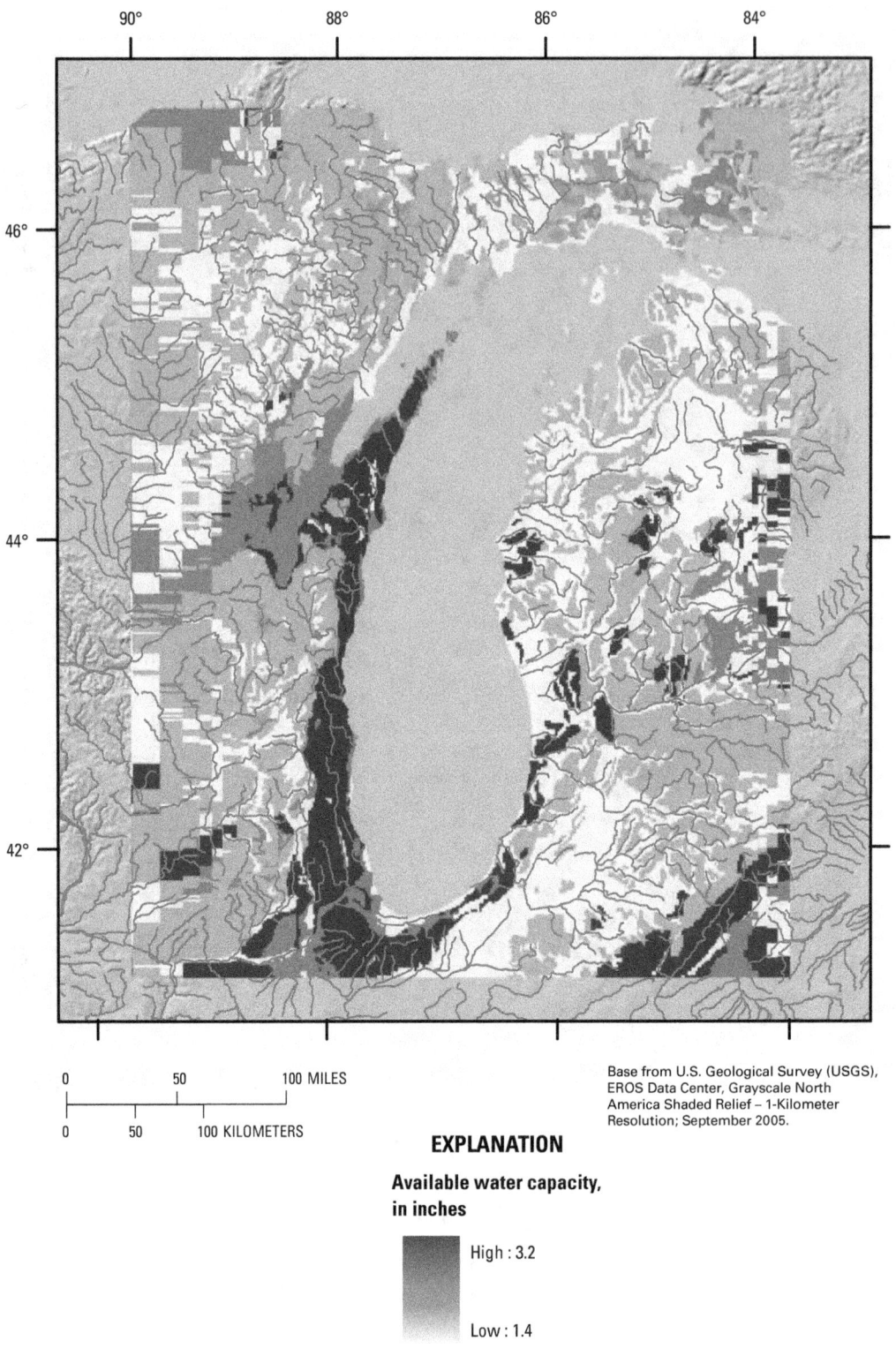

Figure 18. Available water capacity (AWC) for soils in the Lake Michigan Basin test case.

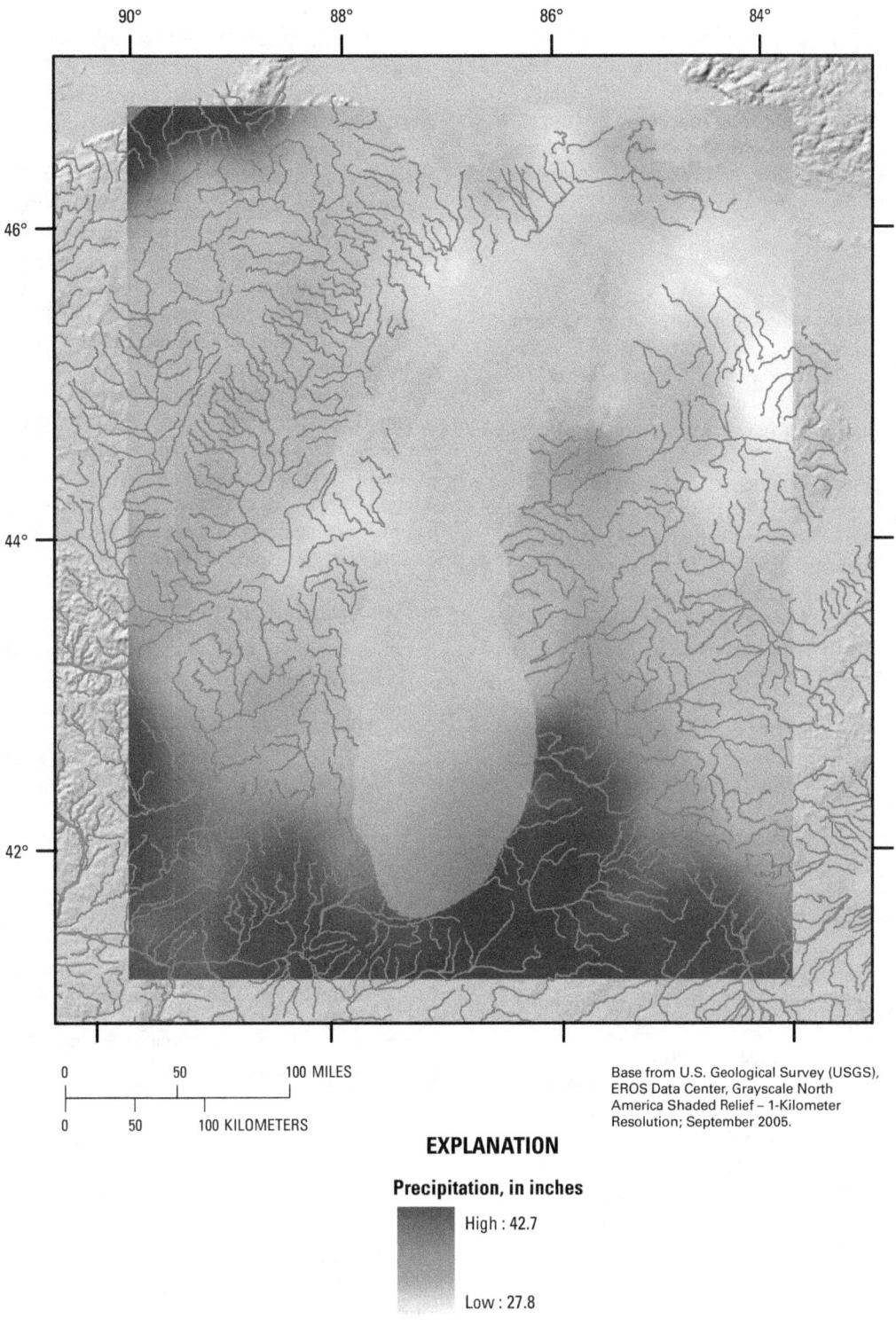

90°   88°   86°   84°

46°

44°

42°

0   50   100 MILES

0   50   100 KILOMETERS

Base from U.S. Geological Survey (USGS),
EROS Data Center, Grayscale North
America Shaded Relief – 1-Kilometer
Resolution; September 2005.

EXPLANATION

Precipitation, in inches

High : 42.7

Low : 27.8

Figure 19.   Mean annual precipitation (1990–2000) for the Lake Michigan Basin test case.

Simulation Details

The SWB model was run for a continuous period extending from 1989 through 2000. Results for 1989 are ignored here, because they represent part of model initialization. The initial snow cover and soil moisture assumed at the beginning of 1989 were assumed to be zero inches and 100 percent of the maximum water capacity, respectively.

Simulation Results

An example of the resulting recharge array is shown in figure 20; annual recharge grids from 1990–2000 were averaged together to produce figure 20. The directive STATS_START_YEAR 1990 was included in the control file in order to exclude the initialization year 1989 from the mean recharge calculation.

Initial SWB model parameters were derived from TR–55 tables (Cronshey and others, 1986) and from the water-holding-capacity tables of Thornthwaite and Mather (1957; see table 10 in this document). SWB model parameter values were iteratively modified in order to increase the correspondence between the distribution of recharge values calculated by the SWB model and the recharge calculated for basins associated with USGS gaging stations.

Figure 21 is a comparison of the SWB estimated recharge to estimates generated through base-flow analysis of streamflow records of various period lengths (Gebert and others, 2007). Recharge estimates from the SWB model were extracted for the contributing area of each of the 100 USGS streamgages within the Lake Michigan Basin for which corresponding base-flow estimates were available. The agreement between the two methods is generally good (slope of regression = 0.94, R^2 = 0.96).

In several instances where the SWB model estimate is as much as twice that of the base-flow analysis, a comparison of the mean annual precipitation for the two periods hints at an explanation for the differences. For example, for the Pike River at Amberg, the SWB code estimates about 6.5 in. of recharge (based on 1990–2000 SWB results), whereas the base-flow analysis suggests that 3.0 in. is more appropriate (based on 1970–1999 gage records). However, comparison of the precipitation records shows that the mean annual precipitation in the 1990–2000 period is almost 4 in. higher at the Pike River site than the mean annual precipitation value for the 1970–2000 period used by Gebert and others (2007).

Model Parameter Sensitivity

Model parameter sensitivity was evaluated by running a customized version of the SWB model in conjunction with PEST parameter-estimation software (Doherty, 2004). PEST calculates model sensitivity to various parameters by varying each parameter value to be tested by a small amount, running the model, and recording the resulting change in model outputs. The sensitivities reported by PEST are composite sensitivities; the values reflect how a fractional change in a given parameter value is translated into change in the model outputs used to make comparisons with the estimates based on analysis of streamflow records. The relative sensitivities shown in figure 22 are obtained by multiplying the composite sensitivities by their respective parameter values (Doherty, 2004). Key parameters are defined and discussed below.

The five parameters with highest relative sensitivities in figure 22 are curve numbers controlling runoff volumes from the most abundant land-use and soil-type combinations. In figure 22, the parameter named cn82_2 refers to the curve number for land use type 82 (row crops) and soil type 2 (loamy till). Of the next six parameters ranking high in terms of relative sensitivity, five relate to correction factors applied to rainfall (rain_corr), snowfall (snow_corr), or evapotranspiration (et_exp, et_slope, et_const). The evapotranspiration parameters that appear in figure 22 are used in the Hargreaves-Samani evapotranspiration calculation method.

Of the remaining parameters shown in figure 22, several are worth pointing out:

- mr1 — Maximum daily recharge to a cell with soil type 1, clay till.

- mr3 — Maximum daily recharge to a cell with soil type 3, fine-textured soils.

- rz82 — Root zone depth of a cell with land use type 82 (row crops).

- rz41 — Root zone depth of a cell with land use type 41 (deciduous forest).

- grow_begin — Day of year in which antecedent runoff condition thresholds and interception change from "dormant" to "growing season" conditions.

- grow_end — Day of year in which antecedent runoff condition thresholds and interception change from "growing season" to "dormant" conditions.

- cfgi_thrs — Value of the continuous frozen ground index above which soils are considered fully frozen, triggering antecedent runoff condition III to be applied in runoff calculations.

The relative sensitivities reported here will be of limited value to SWB applications to other situations. However, it is interesting to see how many of the parameters in this test case are related to evapotranspiration in some way. The rooting-depth parameters in particular seem to deserve some attention in an SWB application; the rooting-depth parameters control the size of the soil-water reservoir from which water may be lost through evapotranspiration.

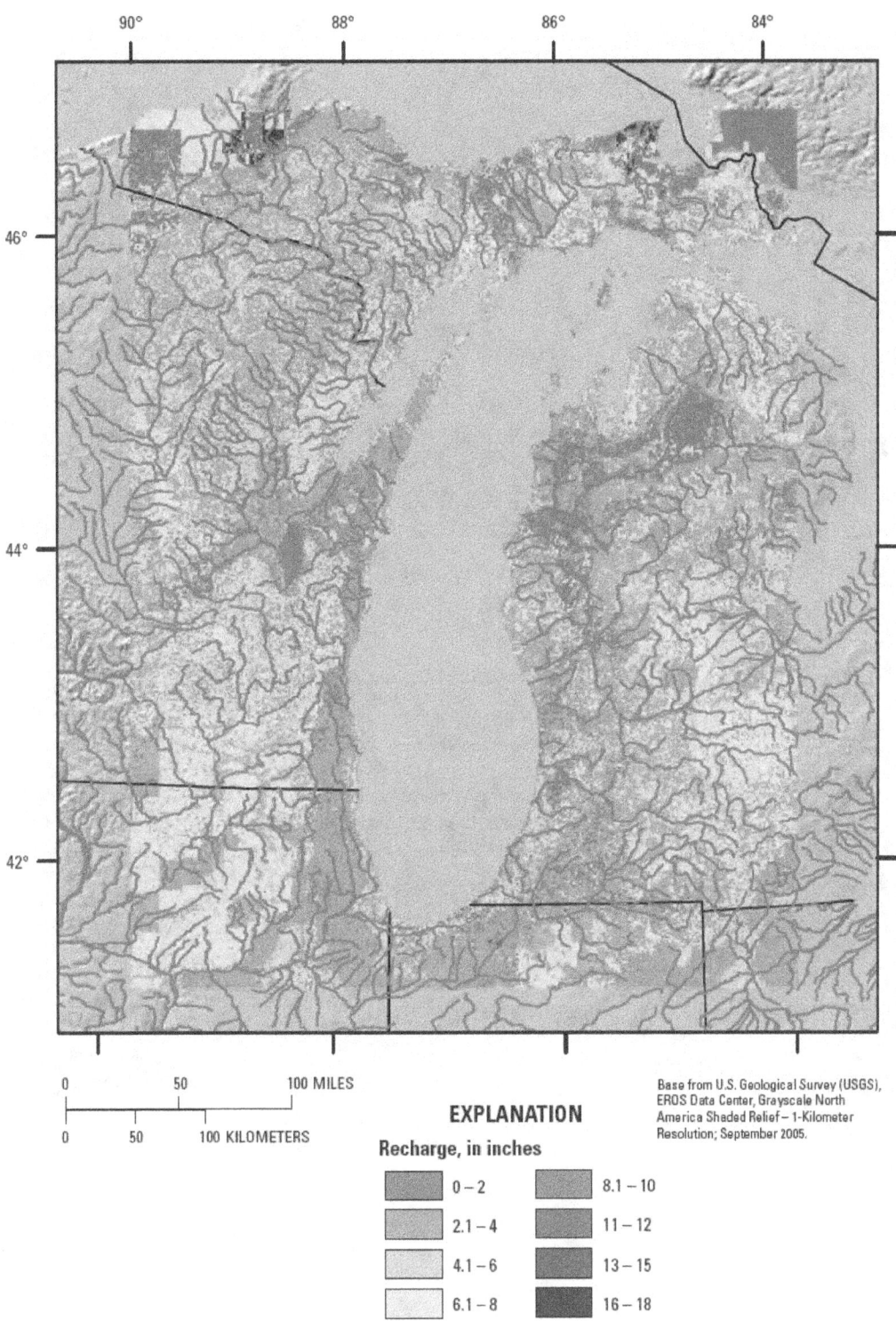

Figure 20.   Example recharge (1990–2000) for the Lake Michigan Basin test case.

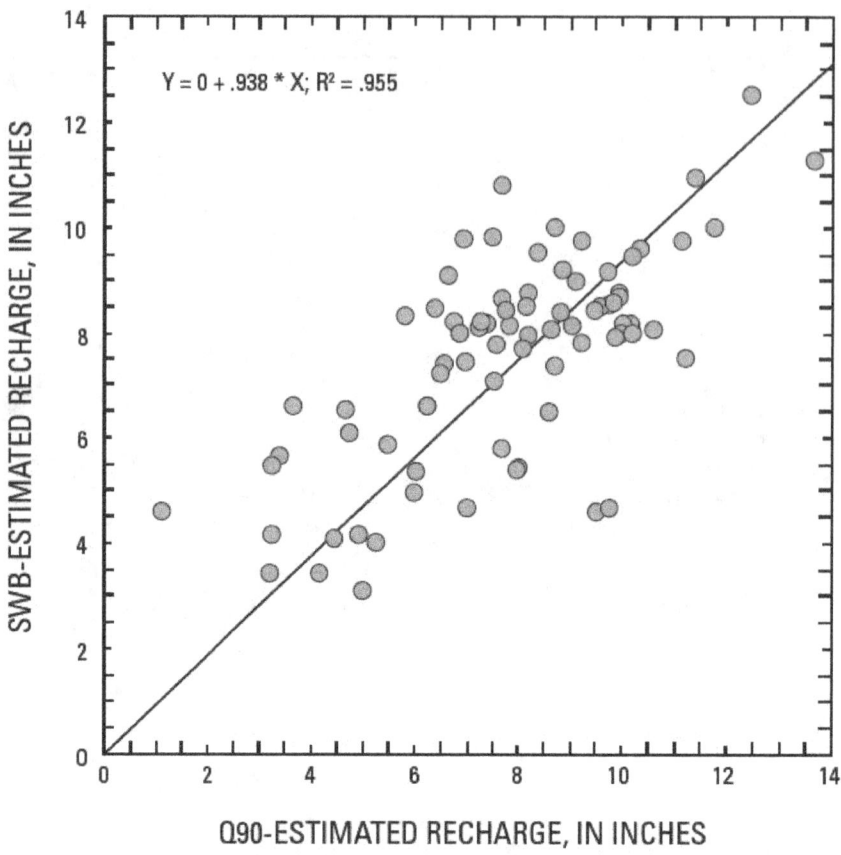

Figure 21.   Comparison of SWB-estimated recharge to Q90-estimated recharge for basins with drainage areas greater than 50 square miles.

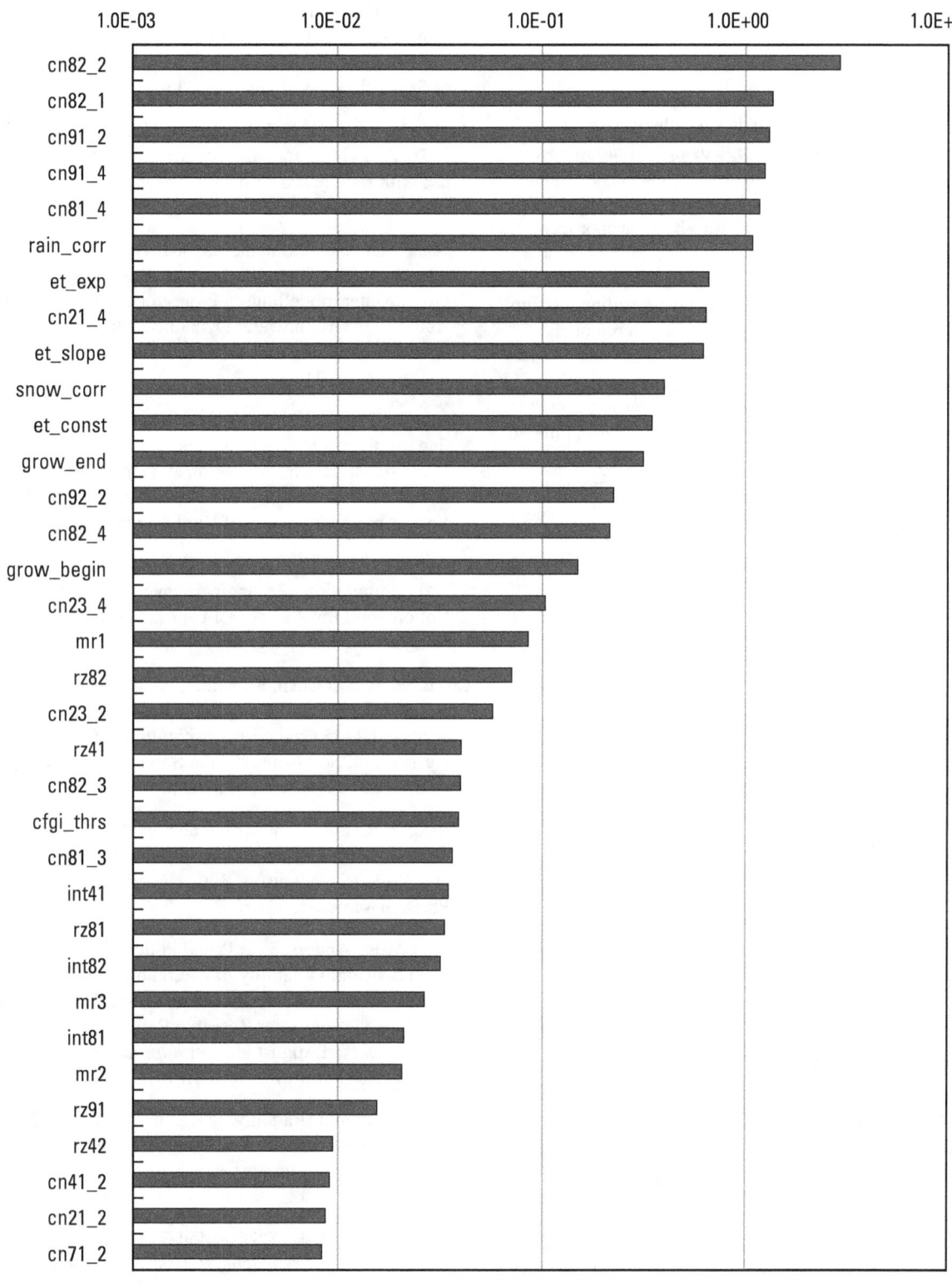

Figure 22. Relative parameter sensitivities for the Lake Michigan Basin SWB model. (Key parameters are defined and discussed in text.)

Summary and Conclusions

A soil-water-balance (SWB) computer code was developed to calculate spatial and temporal variations in groundwater recharge. The soil-water-balance code was designed to calculate these variations in recharge using commonly available geographic information system (GIS) data layers along with tabular climatological data. The code is based on a modified Thornthwaite-Mather soil-moisture-balance approach; components of the soil-moisture balance are calculated on a daily timestep. The soil-moisture calculation is made on a rectangular grid of computational elements; the resulting monthly or annual recharge array can be imported into a regional groundwater-flow model. SWB requires four gridded datasets in order to estimate recharge: (1) hydrologic soil group, (2) land-use/land-cover, (3) available soil-water capacity, and (4) surface-water flow direction. Climate data may be supplied to SWB in either tabular or gridded form.

SWB is designed for application to regional problems rather than site-specific problems. The recharge estimates produced by the SWB code are likely more reliable when averaged over time scales of months to years. Although the code calculates recharge at a daily timestep, there is no consideration of unsaturated-zone flow. In locations where the depth to water table is substantial (more than several meters), there may be a significant lag between the time when SWB generates recharge and the time when that recharge actually reaches the water table.

A test application to the Black Earth Creek Basin shows that the new SWB code is capable of replicating the results generated by the original Visual Basic code (Dripps, 2003). A test application to the Lake Michigan Basin demonstrates the applicability of the technique over a large spatial domain (approximately 116,180 mi^2) and spanning a long time scale (more than 100 years). The use of the SWB code makes it possible to estimate spatially varying transient groundwater-model recharge boundary conditions that reflect observed changes in climate and land use through time.

References Cited

Allen, R.G., Pereira, L.S., Raes, D., and Smith, M., 1998, Crop evapotranspiration (guidelines for computing crop water requirements): Rome, Food and Agriculture Organization of the United Nations, FAO Drainage and Irrigation Paper 56, 300 p.

Allen, R.G., and Pruitt, W.O., 1986, Rational use of the FAO Blaney-Criddle formula: Journal of Irrigation and Drainage Engineering, American Society of Civil Engineers, v. 112, no. 2, p. 139–155.

Akinremi, O.O., McGinn, S.M. and Barr, A.G., 1996, Simulation of soil moisture and other components of the hydrological cycle using a water budget approach: Canadian Journal of Soil Science, v. 76, no. 2, p. 133–142.

Anderson, J.R., Hardy, E.E., Roach, J.T., and Witmer, R.E., 1976, A land use and land cover classification system for use with remote sensor data: U.S. Geological Survey Professional Paper 964, 28 p.

Batelaan, O., and de Smedt, F., 2001, WetSpass—A flexible, GIS-based, distributed recharge methodology for regional groundwater modelling, in Proceedings, Impact of Human Activity on Groundwater Dynamics: Maastricht, Netherlands, International Association of Hydrological Sciences, no. 269, p. 11–18.

Blaney, H.F., and Criddle, W.D., 1966, Determining consumptive use for water developments, in Methods for Estimating Evapotranspiration—Irrigation and Drainage Specialty Conference, November 2–4, Las Vegas, Nev., Proceedings: New York, American Society of Civil Engineers, p. 1–34.

Carroll, R., Pohll, G., Tracy, J., Winter, T., and Smith, R., 2005, Simulation of a semipermanent wetland basin in the Cottonwood Lake area, east-central North Dakota: Journal of Hydrologic Engineering, v. 10, no. 1, p. 70–84.

Cronshey, R., McCuen, R., Miller, N., Rawls, W., Robbins, S., and Woodward, D., 1986, Urban hydrology for small watersheds—TR–55 (2nd ed.): Washington, D.C., U.S. Dept. of Agriculture, Soil Conservation Service, Engineering Division, Technical Release 55, 164 p.

Dripps, W.R., 2003, The spatial and temporal variability of groundwater recharge within the Trout Lake basin of northern Wisconsin: Madison, Wis., University of Wisconsin, Ph.D. dissertation, 231 p.

Dripps, W.R., Anderson, M.P, and Potter, K.W., 2001, Temporal and spatial variability of natural groundwater recharge: Madison, Wis., University of Wisconsin Water Resources Institute, Groundwater Research Report WRI GRR 01–07, 24 p., accessed August 2006 at *http://digital.library.wisc.edu/1711.dl/EcoNatRes.WRIGRR01-07.*

Dripps, W.R., and Bradbury, K.R., 2007, A simple daily soil-water balance model for estimating the spatial and temporal distribution of groundwater recharge in temperate humid areas: Hydrogeology Journal, v. 15, no. 3, p. 433–444.

Doherty, John, 2004, PEST—Model-independent parameter estimation user manual (5th ed.): Brisbane, Australia, Watermark Numerical Computing, [336] p.

Fields Development Team, 2006, Fields—Tools for spatial data: Boulder, Colo., National Center for Atmospheric Research, accessed January 2008 at *http://www.cgd.ucar. edu/Software/Fields.*

Finch, J.W., 2001, Estimating change in direct groundwater recharge using a spatially distributed soil water balance model: Quarterly Journal of Engineering Geology and Hydrogeology, v. 34, no. 1, p. 71–83.

Fullerton, D.S., Bush, C.A., and Pennell, J.N., 2003, Map of surficial deposits and materials in the eastern and central United States (east of 102° West longitude): U.S. Geological Survey Geologic Investigations Series map I–2789, scale 1:2,500,000.

Garen, D.C., and Moore, D.S., 2005, Curve number hydrology in water quality modeling—Uses, abuses, and future directions: Journal of the American Water Resources Association, v. 41, no. 2, p. 377–388.

Gebert, W.A., Radloff, M.J., Considine, E.J., and Kennedy, J.L., 2007, Use of streamflow data to estimate base flow/ ground-water recharge for Wisconsin: Journal of the American Water Resources Association, v. 43, no. 1, p. 220–236.

Green, W.H. and Ampt, C.A., 1911, Studies on soil physics, I. Flow of water and air through soils: Journal of Agricultural Science, v. 4, p. 1–24.

Harbaugh, A.W., Banta, E.R., Hill, M.C., and McDonald, M.C., 2000, MODFLOW-2000, the U.S. Geological Survey modular groundwater model—User guide to modularization processes and the groundwater flow process: U.S. Geological Survey Open-File Report 00–92, 121 p.

Hargreaves, G.H., and Samani, Z.A., 1985, Reference crop evapotranspiration from temperature: Applied Engineering in Agriculture, v. 1, no. 2, p. 96–99.

Hawkins, R.H., 1993, Asymptotic determination of runoff curve numbers from data: Journal of Irrigation and Drainage Engineering, American Society of Civil Engineers, v. 119, no. 2, p. 334–345.

Hjelmfeldt, A.T., Jr., 1991, Investigation of curve number procedure: Journal of Hydraulic Engineering, American Society of Civil Engineers, v. 117, no. 6, p. 725–737.

Jensen, M.E., and Haise, R.H., 1963, Estimating evapotranspiration from solar radiation: Journal of Irrigation and Drainage Division, American Society of Civil Engineers, v. 89, p. 15–41.

Jenson, S., and Domingue, J., 1988, Extracting topographic structure from digital elevation data for geographic information system analysis: Photogrammetric Engineering and Remote Sensing, v. 54, no. 11, p. 1593–1600.

Jyrkama, M.I., and Sykes, J.F., 2007, The impact of climate change on spatially varying groundwater recharge in the Grand River watershed (Ontario): Journal of Hydrology, v. 338, no. 3–4, p. 237–250.

Jyrkama, M.I., Sykes, J.F., and Normani, S.D., 2002, Recharge estimation for transient ground water modeling: Ground Water, v. 40, no. 6, p. 638–648.

Markstrom, S.L., Niswonger, R.G., Regan, R.S., Prudic, D.E., and Barlow, P.M., 2008, GSFLOW—Coupled groundwater and surface-water flow model based on the integration of the precipitation-runoff modeling system (PRMS) and the modular groundwater flow model (MODFLOW–2005): U.S. Geological Survey Techniques and Methods 6–D1, 240 p.

Michels, Helmut, 2007, DISLIN 9.2—A data plotting library: Lindau, Germany, Max Planck Institute for Solar System Research, accessed December 2007 at *http://www.dislin.de/*

Mishra, S.K., and Singh, V.P., 2003, Soil Conservation Service curve number (SCS-CN) methodology: Dordrecht, Netherlands, and Boston, Mass., Kluwer Academic Publishers, Water Science Technology Library, 536 p.

Mockus, Victor, 1964, Estimation of direct runoff from storm rainfall, chap. 10 of National engineering handbook, section 4, hydrology: U.S. Department of Agriculture, Soil Conservation Service, [30] p.

Molnau, Myron, and Bissell, V.C., 1983, A continuous frozen ground index for flood forecasting, *in* Proceedings, Western Snow Conference: Vancouver, Wash., p. 109–119.

Neitsch, S.L., Arnold, J.G., Kiniry, J.R., and Williams, J.R., 2005, Soil and water assessment tool—Theoretical documentation (version 2005): Temple, Tex., Texas Agricultural Experiment Station, [494] p.

Niswonger, R.G., Prudic, D.E., and Regan, R.S., 2006, Documentation of the Unsaturated-Zone Flow (UZF1) Package for modeling unsaturated flow between the land surface and the water table with MODFLOW–2005: U.S. Geological Survey Techniques and Methods 6–A19, 74 p.

O'Callaghan, J.F., and Mark, D.M., 1984, The extraction of drainage networks from digital elevation data: Computer Vision, Graphics, and Image Processing, v. 28, no. 3, p. 323–344.

R Development Core Team, 2008, R, a language and environment for statistical computing—reference index: Vienna, Austria, R Foundation for Statistical Computing, accessed April 2008 at *http://www.R-project.org/*

Scanlon, B.R., Healy, R.W., and Cook, P.G., 2002, Choosing appropriate techniques for quantifying groundwater recharge: Hydrogeology Journal, v. 10, no. 1, p. 18–39.

Schroeder, P.R., Dozier, T.S., Zappi, P.A., McEnroe, B.M., Sjostrom, J.W., and Peyton, R.L., 1994, The hydrologic evaluation of landfill performance (HELP) model—Engineering documentation for version 3: Washington, D.C., U.S. Environmental Protection Agency Office of Research and Development, EPA/600/R–94/168b, [126] p.

Steenhuis, T.S., and Van der Molen, W.H., 1986, The Thornthwaite-Mather procedure as a simple engineering method to predict recharge: Journal of Hydrology, v. 84, no. 3–4, p. 221–229.

Thornthwaite, C.W., 1948, An approach toward a rational classification of climate: Geographical Review, v. 38, no. 1, p. 55–94.

Thornthwaite, C.W., and Mather, J.R., 1957, Instructions and tables for computing potential evapotranspiration and the water balance: Centerton, N.J., Laboratory of Climatology, Publications in Climatology, v. 10, no. 3, p. 185–311.

Thornthwaite, C.W., and Mather, J.R., 1955, The water balance: Centerton, N.J., Laboratory of Climatology, Publications in Climatology, v. 8, no. 1, p. 1–104.

Turc, L., 1961, Evaluation des besoins en eau d'irrigation, évapotranspiration potentielle, formule climatique simplifée et mise à jour (In French; original unseen): Annales Agronomiques, v. 12 no. 1, p. 13–49.

U.S. Department of Agriculture, Natural Resources Conservation Service, 2006, Digital general soil map of U.S.: accessed January 2007 at *http://SoilDataMart.nrcs.usda.gov/*.

U.S. Geological Survey, 2000a, U.S. GeoData Digital Elevation Models: U.S. Geological Survey Fact Sheet 040–00, accessed January 2007 at *http://erg.usgs.gov/isb/pubs/factsheets/fs04000.html*.

U.S. Geological Survey, 2000b, National Land Cover Dataset: U.S. Geological Survey Fact Sheet 108–00, accessed January 2007 at *http://erg.usgs.gov/isb/pubs/factsheets/fs10800.html*.

University of Wisconsin Land Information and Computer Graphics Facility, 1988, Dane County digital soils map: Madison, Wis., Dane County Land Information Office.

Vörösmarty, C.J., Federer, C.A., and Schloss, A.L., 1998, Potential evaporation functions compared on US watersheds—Possible implications for global-scale water balance and terrestrial ecosystem modeling: Journal of Hydrology, v. 207, no. 3–4, p. 147–169.

Woodward, D.E., Hawkins, R.H., Hjelmfelt, A.T., Jr., Van Mullem, J.A., and Quan, Q.D., 2002, Curve number method—Origins, applications and limitations: U.S. Department of Agriculture, Natural Resources Conservation Service, Hydraulics and Hydrology Technical Reference, 10 p., accessed May 2008 at *http://www.wsi.nrcs.usda.gov/products/W2Q/H&H/docs/H&H_papers/curve_number/CN_origins.doc*.

Woodward, D.E., Hawkins, R.H., Jiang, R., Hjelmfelt, A.T., Van Mullem, J.A., and Quan, Q.D., 2003, Runoff curve number method—Examination of the initial abstraction ratio, *in* World Water and Environmental Resources Congress 2003, Conference Proceedings Paper, June 24–26, Philadelphia: American Society of Civil Engineers, [12] p.

Willmott, C.J., 1977, WATBUG—A FORTRAN IV algorithm for calculating the climatic water budget: Elmer, N.J., Laboratory of Climatology, Publications in Climatology, v. 30, no. 2., p. 1–55.

Wisconsin Department of Natural Resources, 1998, WISCLAND land cover (WLCGW930): Madison, Wis., accessed May 2007 at *http://dnr.wi.gov/maps/gis/datalandcover.html*.

Appendixes 1 and 2

Appendix 1: SWB Module Description

This section describes the function of each of the program modules that make up the SWB code.

Main Program and Module

main f95 — The program main f95 is a short program with the sole purpose of calling subroutine model_Run, which is contained in module model f95. All other modules within the SWB code are implemented as subroutines.

model.f95 — This subroutine contains two primary subroutines along with several support subroutines. The two primary subroutines are model_Run, and model_Solve. The model_Run subroutine reads and interprets the directives included in the control file. Once the control file has been read in, the subroutine model_Solve is called to perform the simulation.

Support Modules

Code that is intended for general use by one or more other modules has been placed into separate support modules, as described below.

types f95 — The types f95 module defines reusable data types for manipulating whole model grids and individual grid cells, as well as numerous support functions and global parameters that may be used by any program module.

grid.f95 — The grid.f95 module contains subroutines that instruct the program in creating, reading, and destroying Arc ASCII and Surfer grids.

stats f95 — The stats.f95 module defines data structures and subroutines that compute basic statistics for each model state variable on a daily, monthly, or annual timestep.

RLE f95 — This module takes grid output from an SWB variable and reads or writes the gridded values in the form of unformatted Fortran binary files. While reading and writing, the subroutine implements "run length encoding," which compresses the size of the output file on the basis of the level of redundancy in the gridded data set.

climatological_functions.f95 — This module contains functions used in support of potential evapotranspiration calculations. In particular, this module contains functions to calculate or estimate extraterrestrial and clear-sky radiation, sunset angle, number of daylight hours for a given day of the year, and other similar functions.

Process Modules

Each process module encapsulates methods used to calculate specific pieces of the water balance. The architecture of the code makes it relatively simple to add new process modules if they become necessary.

Evapotranspiration

Five methods of estimating the potential evapotranspiration are included in this version of the code. Of the five methods, only the Hargreaves-Samani method is adapted to computations using gridded precipitation and temperature data.

et_blaney_criddle.f95 — Computes potential evapotranspiration by use of a modified FAO 24 Blaney-Criddle method (Allen and Pruitt, 1986). The method requires data on mean air temperature, minimum relative humidity, percentage of possible sunshine hours, and wind speed.

et_jensen_haise.f95 — Computes potential evapotranspiration by use of the Jensen and Haise method (1963). Solar radiation is estimated by means of the Angstrom formula, based on the percentage of actual sunshine hours relative to the number of hours between sunrise and sunset. Data required include percentage of possible sunshine hours and mean air temperature.

et_turc.f95 — Computes potential evapotranspiration by use of the Turc method (1961). Solar radiation is estimated by means of the Angstrom formula, on the basis of the percentage of actual sunshine hours relative to the number of hours between sunrise and sunset. Data required include percentage of possible sunshine hours, mean relative humidity, and mean air temperature.

et_thornthwaite_mather f95 — Computes potential evapotranspiration by use of the Thornthwaite and Mather method (1957). The method requires only daily mean air temperature data.

et_hargreaves f95 — Computes potential evapotranspiration by use of the Hargreaves-Samani method (Hargreaves and Samani, 1985; Allen and others, 1998). The method requires data on mean, maximum, and minimum air temperature.

Runoff

runoff_curve_number.f95 — Computes runoff from an individual grid cell by use of the NRCS curve number method.

Soil Moisture

sm_thornthwaite_mather.f95 — Computes soil moisture on the basis of the procedure and tables included in Thornthwaite and Mather (1955, 1957).

Appendix 2: Preparation of Gridded Climatological Data for SWB

It is beyond the scope of this report to detail all steps involved in creating daily precipitation and temperature grids; however, this section outlines an abbreviated description of the method used in the test case.

Climatological Data Source

Daily surface observation data were obtained from the National Climatic Data Center, a branch of the National Oceanic and Atmospheric Administration. The URL (as of December 2007) for these data is *http://cdo.ncdc.noaa.gov/pls/plclimprod/poemain.accessrouter?datasetabbv=SOD*.

In the text that follows, it is assumed that climatological values from all stations within a given meteorological division are downloaded from the National Climatic Data Center. Because of band-width limitations, each downloaded file will likely cover 10–20 years of the entire period of record in a series of files corresponding to a subset of the period of record. The resulting files contain one line of comma-delimited data for each month of station operation; the file will contain data from all stations within the climatological division. The data format is fully documented in a National Climatic Data Center publication, which may be found (as of December 2007) at the following URL:

http://cdo.ncdc.noaa.gov/cdo/soddoc.txt

An example of part of one of these files is shown below.

```
DSET,COOPID,WBNID,STATION NAME,CD,ELEM,UN,YEARMO,DAHR, DAY01,F,F,…

----,------,-----,--------------------,--,----,--,------,----,------,-,-,…

3206,330862,99999,BOWLING GREEN WWTP,99,PRCP,HI,190001,0199, 00000,T,1,…

3206,330862,99999,BOWLING GREEN WWTP,99,SNOW,TI,190001,0199, 00000,T,1,…

3206,330862,99999,BOWLING GREEN WWTP,99,TMAX, F,190001,0199, 00017, ,1,…
```

A simple bash script that will create a separate tab-delimited data file for each station for the available period of record is shown below. Note that the script makes use of the Free Software Foundation's GNU Core Utilities (*http://www.gnu.org/software/coreutils/*) as distributed by the Cygwin Project (*http://www.cygwin.com*).

```
#!/bin/bash
# script requires two command-line arguments:
# a two-character state abbreviation code, and the climatological
# data type (e.g. "PRCP, "SNOW")
#
# script also assumes that all downloaded data files are saved with
# a ".csv" file extension

# assemble file for entire period of record from intermediate files,
# stripping all header information from the files in the process
for i in $(ls $1*.csv); do
  egrep -v '(----|COOPID)' $i >> $1_precip.csv
  echo Item: $i
done

# sort the resulting file by STATION NAME
sort -t ',' -k 4,4 $1_precip.csv > $1_precip_sort.csv

# create a list of the unique station names contained in the sorted file
# also trim whitespace from the end of the station name and replace
# spaces with underscore characters
cut -d ',' -f 4 $1_precip_sort.csv| uniq | sed 's/[ \t]*$//' | tr ' ' '_' >
$1_ST_NAMES.lst

# for each station name, output a file of tab-delimited values sorted by
date
for i in $(cat $1_ST_NAMES.lst); do
    ONAME=$(echo ${i//_/ })
    FNAME=$(echo $1)_$(echo $i)_$(echo $2).txt
    egrep "$ONAME.*$2" $1_precip_sort.csv | tr ',' '\t' | sort -k 8,8 >
$FNAME
done
```

Once files for individual sites are created, a Perl script is used to filter the data from each file on the basis of date and time interval of interest and on whether valid data existed at all for the site. The Perl script collects the daily values for each site and writes these values to a file for further processing, described below. The most important thing the script does is add location information to the header and rearrange the data in columnar format as required by the R statistical package (R Development Core Team, 2008).

Performing Interpolations

A wide range of software tools and techniques may be used to create interpolated data grids suitable for use as daily input for the SWB model. In our example, a file containing the x-coordinate, y-coordinate, and precipitation or temperature value was created for each day of simulation. An interpolation was performed for each of these XYZ files, and an Arc ASCII grid file was written to disk. In this application, an R package called Fields (*http://www.image.ucar. edu/GSP/Software/Fields/*) was used to perform a thin-plate spline on the precipitation data.

```
# Example R script to perform thin plate spline on precip data

# requires the "Fields" library
library(fields)

# create x, y, and z values (simulated rainfall data)
# in real application we would read these from a file

x<-runif(20)*10000
y<-runif(20)*20000
z<-pmax((runif(20)-0.5)*3,0)

# perform the thin plate spline fit
fit<-Tps(cbind(x,y),z)

# create the list of points for which we would like interpolated data
x0<-seq(min(x),max(x),100)
y0<-seq(min(y),max(y),100)
grid.l<-list(X=x0,Y=y0)

# create the matrix of interpolated results
outp<-predict.surface(fit,grid.list=grid.l,extrap=T)

# plot up results
surface(outp)
```

Figure 2–1 shows the surface resulting from the code listed above.

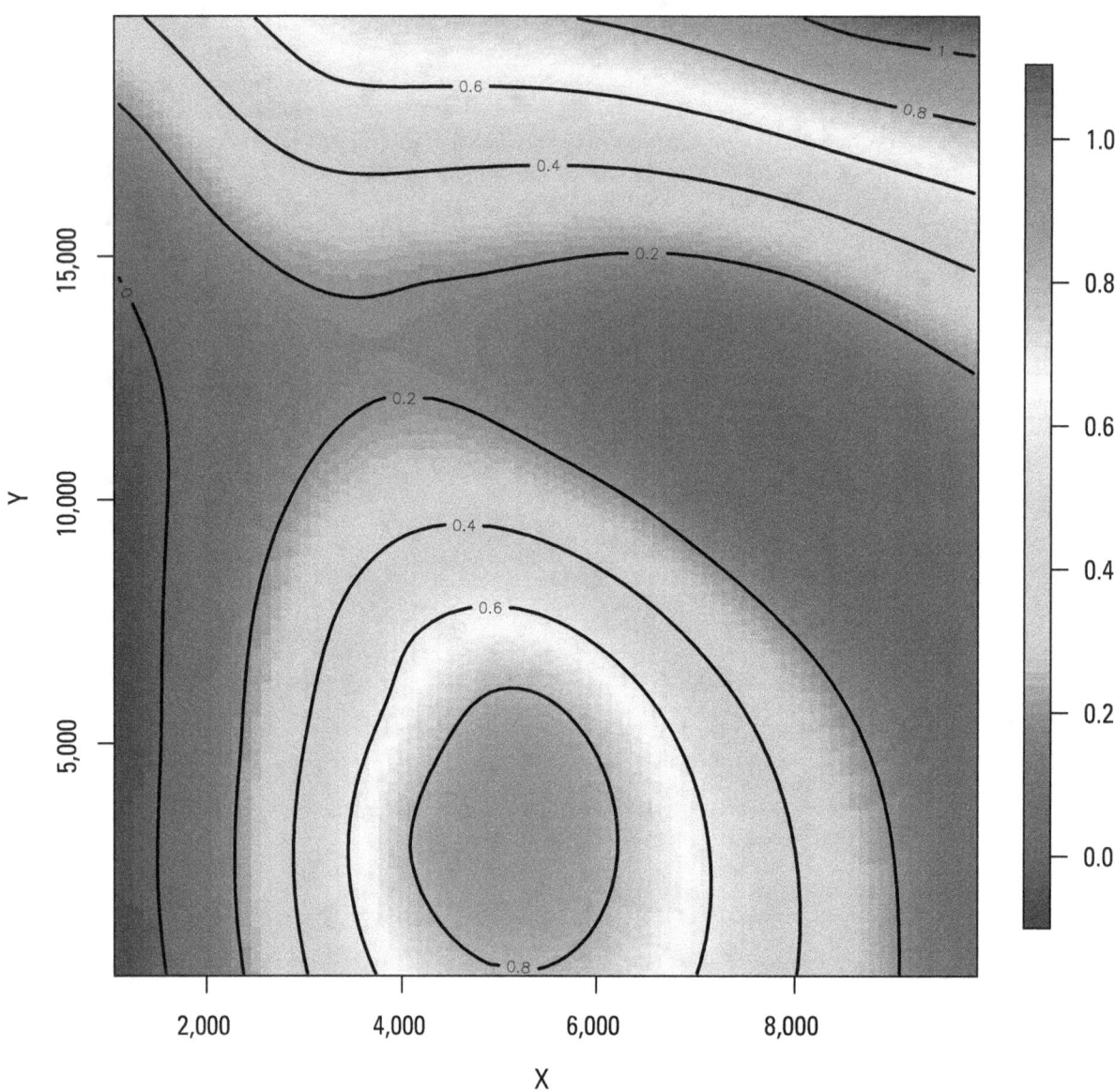

Figure 2–1.   Example of the output from the thin plate spline technique in the fields library of the R statistical package.

Fortunately, the Arc ASCII grid file format is relatively simple to create. Within R, a trivial function may be written to reformat the interpolated values and write them out in an Arc ASCII file format. Example code is listed below.

```
#define a function to write Arc ASCII grid files
writeArcASCII<- function(fname,matvar,llx,lly,nx,ny,grid.delta) {
  cat(paste("NCOLSz",nx,"\n"),file=fname)
  cat(paste("NROWS",ny,"\n"),file=fname,append=T)
  cat(paste("XLLCORNER",llx,"\n"),file=fname,append=T)
  cat(paste("YLLCORNER",lly,"\n"),file=fname,append=T)
  cat(paste("CELLSIZE",grid.delta,"\n"),file=fname,append=T)
  cat(paste("NODATA_VALUE -99999","\n"),file=fname,append=T)

  for(i in ny:1) {
    cat(matvar[,i],file=fname,append=T)
    cat("\n",file=fname,append=T)

  }
}
```